T0331710

Introduction to the Theory of Coherence
and Polarization of Light

All optical fields undergo random fluctuations. They may be small, as in the output of many lasers, or they may be appreciably larger as in light generated by thermal sources. The underlying theory of fluctuating optical fields is known as coherence theory. An important manifestation of the fluctuations is the phenomenon of partial polarization. Actually, coherence theory deals with appreciably more than with fluctuations. Unlike usual treatments it describes optical fields in terms of observable quantities and elucidates how such quantities, for example, the spectrum of light, change as light propagates.

This book is the first to provide a unified treatment of the phenomena of coherence and polarization. The unification has been made possible by very recent discoveries, largely due to the author of this book.

The subjects treated in this volume are of considerable importance for graduate students and for research workers in physics and in engineering who are concerned with optical communications, with propagation of laser beams through fibers and through the turbulent atmosphere, with optical image formation, particularly in microscopes, and with medical diagnostics, for example. Each chapter contains problems to aid self-study.

EMIL WOLF is Wilson Professor of Optical Physics at the University of Rochester, and is renowned for his work in physical optics. He has received many awards, including the Ives Medal of the Optical Society of America, the Albert A. Michelson Medal of the Franklin Institute and the Marconi Medal of the Italian Research Council. He is the recipient of seven honorary degrees from universities around the world. Professor Wolf co-authored the well-known text *Principles of Optics* (with Max Born, seventh edition, Cambridge University Press, 1999) and *Optical Coherence and Quantum Optics* (with Leonard Mandel, Cambridge University Press, 1995). He has also been the editor of a well-known series *Progress in Optics* since its inception. Fifty volumes of *Progress in Optics* have been published to date.

Introduction to the
Theory of Coherence and
Polarization of Light

EMIL WOLF

Wilson Professor of Optical Physics
University of Rochester, Rochester, NY 14627, USA
and
Provost's Distinguished Research Professor
CREOL & FPCE, College of Optics and Photonics, University of Central Florida,
Orlando, FL 32816, USA

CAMBRIDGE
UNIVERSITY PRESS

CAMBRIDGE
UNIVERSITY PRESS

Shaftesbury Road, Cambridge CB2 8EA, United Kingdom

One Liberty Plaza, 20th Floor, New York, NY 10006, USA

477 Williamstown Road, Port Melbourne, VIC 3207, Australia

314–321, 3rd Floor, Plot 3, Splendor Forum, Jasola District Centre, New Delhi – 110025, India

103 Penang Road, #05–06/07, Visioncrest Commercial, Singapore 238467

Cambridge University Press is part of Cambridge University Press & Assessment,
a department of the University of Cambridge.

We share the University's mission to contribute to society through the pursuit of
education, learning and research at the highest international levels of excellence.

www.cambridge.org
Information on this title: www.cambridge.org/9780521822114

First published 2007

A catalogue record for this publication is available from the British Library

ISBN   978-0-521-82211-4   Hardback

Dedicated to the memory of Leonard Mandel,
a dear friend and colleague of more than forty years.

Thomas Young[1]
(1773–1829)

Gabriel Stokes[2]
(1819–1903)

Frits Zernike[3]
(1888–1966)

The three pioneers who laid the foundations of the theory
presented in this book

Contents

Preface

> *". . . the image that will be formed in a photographic camera – i.e. the distribution of intensity on the sensitive layer – is present in an invisible, mysterious way in the aperture of the lens, where the intensity is equal at all points."*
> F. Zernike, discussing coherence in a lecture published in *Proc. Phys. Soc.* (London), **61** (1948), 158.

The optical coherence and polarization phenomena with which this book is concerned have their origin in the fact that all optical fields, whether encountered in nature or generated in a laboratory, have some random fluctuations associated with them. The monochromatic sources and monochromatic fields discussed in most optics textbooks are not encountered in real life.

For thermal light, the fluctuations are mainly due to spontaneous emission of radiation from atoms which generate the field. For laser light the fluctuations are due to uncontrollable causes such as mechanical vibrations of the mirrors at the end of the cavity, temperature fluctuations and, again, arise also because contributions from spontaneous emission are always present. For a well-stabilized laser these effects are largely manifested in phase fluctuations rather than in amplitude fluctuations and also in the chaotic behavior of the laser output that may be detected over a sufficiently long time. In the optical region of the electromagnetic spectrum, the field fluctuations are too rapid to be directly measurable. The theory of coherence and polarization involves average quantities that can be measured. Consequently, the theory deals with observable quantities.

When I started writing this book my intention was to provide an introductory text on the subject of classical optical coherence theory alone. Although there are now several books and book chapters devoted to coherence, none of them seems to me to treat the subject at a level appropriate for a reader who has a reasonable knowledge of elementary optics and none presents the basic concepts and results of coherence theory in a sufficiently sound and yet not too abstract manner. By the time I had written several chapters, a new development had taken place, namely the discovery that coherence and polarization of light are two aspects of statistical optics which are intimately related and can be treated in a unified manner. Until then coherence and polarization had been considered as being essentially independent of each other – the only apparent link was the term "coherency matrix," a 2 × 2 correlation matrix which has been used in the analysis of elementary polarization problems since the 1930s. This term is actually a misnomer, because such a matrix has nothing to do with coherence, as the term is now understood. Coherence is essentially a consequence of correlations between some components of the fluctuating electric field at two (or more) points and is manifested by the sharpness of fringes in Young's interference experiment. Polarization, on the other hand, is a manifestation of correlations involving components of the fluctuating electric field at a single point and may be determined with the help of polarizers, rotators and phase plates. Both concepts reflect "degrees of order" in an electromagnetic field, but they pertain to somewhat different statistical aspects of it. The theories of coherence and polarization are, however, concerned not only with order and disorder in optical fields. Their basic tools are correlation functions and correlation matrices which, unlike some directly measurable quantities such as the spectrum of light, obey precise propagation laws. With the help of these laws one may determine, for example, spectral and polarization changes as light propagates, whether in free space or in a medium, which may be either deterministic (e.g. a glass fiber) or random (e.g. the turbulent atmosphere). Consequences of these laws are among the most useful aspects of the theory.

Until very recently, coherence phenomena have been studied largely on the basis of scalar theory, whereas polarization requires a vector treatment. It was actually a generalization of the concept of coherence from scalar fields to electromagnetic vector fields, introduced only a few years ago, that has made it evident that coherence and polarization of light, whilst distinct phenomena, are just two closely related aspects of statistical optics; and that many features of fluctuating electromagnetic fields can be fully understood only when they are treated in close partnership.[1] This discovery has not only enriched both subjects, but also has already provided new insights into many aspects of statistical optics. This development, which is discussed in the concluding chapter, is likely to find useful applications, for example, in connection with optical communication, with imaging by laser radar and in medical diagnostics, but undoubtedly other applications will be forthcoming.

In order to provide a treatment of the subject which is not too demanding mathematically and which will help the reader to acquire a working knowledge of it, detailed proofs are sometimes omitted. Most of them can be found in M. Born and E. Wolf, *Principles of Optics* (Cambridge University Press, Cambridge, 7th (expanded) edition, 1999) and in L. Mandel and E. Wolf, *Optical Coherence and Quantum Optics* (Cambridge University Press,

---

[1] Developments leading to the recognition that there exists an intimate relationship between the phenomena of coherence and polarization are discussed in an article by E. Wolf, "Young's interference experiment and its influence on the development of statistical optics" in volume 50 of *Progress in Optics* (Amsterdam, Elsevier, 2007).

Cambridge, 1995),[1] which also contain more detailed accounts of the subject, with full references. The historical development of coherence theory is outlined in an article in *Selected Works of Emil Wolf with Commentary* (World Scientific, Singapore, 2001), pp. 620–633. Accounts of the development of the theory of polarization can be found in E. Collett, *Polarized Light* (M. Dekker, New York, 1993) and in C. Brosseau, *Fundamentals of Polarized Light* (J. Wiley, New York, 1998).

Some readers may note that parts of the presentation resemble fairly closely the treatments given in B&W and M&W. This is mainly due to the fact that I had difficulties in providing a different formulation, but it should be clear that this book uses more elementary and less rigorous analysis, aimed at non-specialists, especially teachers and students, who might perhaps also find it helpful that problems are included at the end of each chapter. Additional problems can be found in M&W.

I am grateful to Dr. Gale Gant and Dr. Don Nicolson for the photograph of the 20-foot Michelson stellar interferometer at Mount Wilson Observatory, built in the 1920s. The photograph, reproduced as Fig. 3.12, was taken around the year 2000.

In writing this book I have greatly benefited from the assistance of many colleagues and students who read drafts of the manuscript and helped in weeding out errors and improving the text. I would particularly like to acknowledge substantial advice that I received from Professor Taco Visser and also useful suggestions from Prof. Jannick Rolland, Mrs. Nicole Carlson-Moore, Dr. David Fischer, Dr. Olga Korotkova, Dr. Mircea Mujat, Mr. Jonathan Petruccelli, Mr. Mohamed Salem, Mr. Thijs Stegeman, Dr. Tomohiro Shirai, Mr. Mayukh Lahiri and Mr. Thomas van Dijk. I am also obliged to Mr. Mohamed Salem and Dr. Sergei Volkov for help with checking the proofs.

The staff of the Physics–Optics–Astronomy Library of the University of Rochester provided much help, especially with locating articles and checking references. I am much obliged to Mrs. Patricia Sulouff, the Head Librarian, and to Mrs. Sandra Cherin and Mrs. Miriam Margala for their assistance.

I am greatly obliged to Dr. Greg Gbur for preparing the excellent line drawings of most of the figures and also for the beautiful figure which appears on the front cover.

Some of the research described in this book, especially that connected with the unified theory of coherence and polarization discussed in Chapter 9, was supported by the Air Force Office of Scientific Research (AFOSR). I am much indebted to Dr. Arje Nachman of AFOSR for his continued support over many years and for his interest in our work.

I acknowledge, with many thanks, the very substantial help provided by my secretary, Mrs. Ellen Calkins, who, without any complaints, typed and re-typed numerous versions of the manuscript and also prepared the author index.

I wish to express my appreciation to my patient wife, Marlies, who spent long periods in solitude whilst I was preparing the manuscript.

I presented much of the material contained in this book in courses to Physics and Optics graduate students at the University of Rochester and at the University of Central Florida;

---

[1] References to these books are abbreviated in the present work as B&W and M&W, respectively.

but a good part of the text is an expanded version of notes that I prepared for a course which I taught for many years at Annual Meetings of the Optical Society of America. I am indebted to Dr. Simon Capelin, the Publishing Director for Physical Sciences of Cambridge University Press, for suggesting that I expand the notes into a book and for encouraging me to do so.

Finally I wish to express my appreciation to the staff of Cambridge University Press and, particularly, to J. Bottrill, the production editor, K. Howe, the production manager and to Dr. S. Holt who copy-edited the manuscript, for their cooperation and for having transformed an imperfect manuscript into a beautiful end product.

Department of Physics and Astronomy   Emil Wolf
University of Rochester
Rochester, NY 14627, USA

Spring 2007

1

Elementary coherence phenomena

1.1 Interference and statistical similarity

The simplest manifestation of coherence between light vibrations at different points in an optical field is provided by the phenomenon of interference. In fact, as we will learn later, some features of the interference pattern provide a quantitative measure of the coherence between light vibrations at two points in space and at the two instants of time.

Let us first consider light vibrations at a point P in an optical field. For the sake of simplicity we will ignore, to begin with, the polarization properties of light and we may then represent the light vibrations at a point in the field by a scalar, say $U(t)$. If the light were monochromatic, it would be expressed as

$$U(t) = a \cos(\phi - \omega t), \tag{1}$$

where a and ϕ are the (constant) amplitude and phase, respectively, ω is the frequency and t denotes the time. However, as we have already noted, monochromatic light is an idealization which is never encountered in nature or in a laboratory. Light that in some respects imitates monochromatic light most closely is so-called *quasi-monochromatic* light. It is defined by the property that its effective bandwidth, $\Delta \omega$ is much smaller than its mean frequency $\bar{\omega}$, i.e. that

$$\frac{\Delta \omega}{\bar{\omega}} \ll 1. \tag{2}$$

For such light the amplitude and the phase are no longer constant and its vibrations at a point in space may be represented by a generalization of Eq. (1), viz.,

$$U(t) = a(t)\cos[\phi(t) - \bar{\omega}t], \tag{3}$$

where the amplitude $a(t)$ and the phase $\phi(t)$ now depend on time, and generally fluctuate randomly. With the help of elementary Fourier analysis one can show (M&W, Section 3.1, especially pp. 99–100) that for quasi-monochromatic light $a(t)$ and $\phi(t)$ will vary very

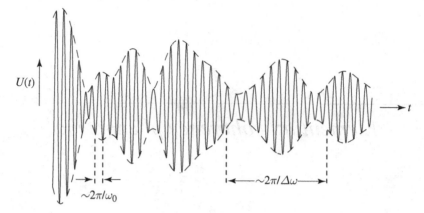

Fig. 1.1 Illustrating the behavior of the light vibrations $U(t)$ of quasi-monochromatic light at a point in space.

slowly over any time interval Δt which is short compared with the reciprocal bandwidth of the light, i.e. for time intervals

$$\Delta t \ll \frac{2\pi}{\Delta\omega}, \tag{4}$$

as illustrated in Fig. 1.1, where $a(t)$ is indicated by dashed lines.

We have implicitly assumed that we are dealing with a "steady-state" field. By this it is meant that the statistical behavior that underlies the fluctuations of the field does not change over the course of time. In the language of the statistical theory which we will briefly consider later (see Section 2.1), such a field is said to be *statistically stationary*.

Let us now consider vibrations $U_1(t)$ and $U_2(t)$ at two points P_1 and P_2 in a quasi-monochromatic field:

$$U_1(t) = a_1(t)\cos[\phi_1(t) - \bar{\omega}t], \tag{5a}$$

$$U_2(t) = a_2(t)\cos[\phi_2(t) - \bar{\omega}t]. \tag{5b}$$

Suppose that we superpose these vibrations at another point P, for example by placing an opaque screen across the field, with pinholes at P_1 and P_2. Apart from some unessential geometrical factors which depend on the size of the pinholes and on the angles of incidence and diffraction, which we will assume to be small, the vibrations at the point of superposition may be represented by the formula

$$U(t) = U_1(t) + U_2(t)$$

$$= a_1(t)\cos[\phi_1(t) - \bar{\omega}t] + a_2(t)\cos[\phi_2(t) - \bar{\omega}t + \delta] \tag{6}$$

where δ is the phase difference introduced between the two beams propagating from P_1 to P and from P_2 to P, respectively.

The instantaneous intensity $I(t)$ at the point P may be defined, in appropriate units, as the square of $U(t)$:

$$I(t) = U^2(t)$$
$$= I_1(t) + I_2(t) + I_{12}(t). \tag{7}$$

where

$$I_1(t) = a_1^2(t)\cos^2[\phi_1(t) - \bar{\omega}t], \tag{8a}$$

$$I_2(t) = a_2^2(t)\cos^2[\phi_2(t) - \bar{\omega}t + \delta], \tag{8b}$$

and

$$I_{12}(t) = a_1(t)a_2(t)\{\cos[\phi_1(t) + \phi_2(t) - 2\bar{\omega}t + \delta] + \cos[\phi_1(t) - \phi_2(t) - \delta]\}. \tag{8c}$$

The expression (8c) follows from an elementary trigonometric identity.

Because of the very rapid fluctuations of optical fields, one can never measure the instantaneous intensity but only its average over some time interval $-T \le t < T$ which is large compared with the reciprocal bandwidth of the light, i.e. $T \gg 1/\Delta\omega$. To avoid ambiguity, and also for reasons which will become evident later when we speak of ergodicity (Section 2.1), we will formally let $T \to \infty$ and define the time-averaged intensity by the formula

$$\langle I \rangle_t = \lim_{T \to \infty} \frac{1}{2T} \int_{-T}^{T} I(t)\mathrm{d}t. \tag{9}$$

We attached the suffix t on the angular brackets to stress that the brackets refer to a time average and also to distinguish it from another type of average which we will encounter later.

Suppose that the amplitudes of the field at the points P_1 and P_2 are effectively independent of time (as is approximately true for the output of a well-stabilized single-mode laser) and also that they are equal to each other, i.e. that $a_1(t) = a_2(t) = a = $ constant. Then, on taking the time average, we obtain from Eqs. (7) and (8) the following expression for the average intensity at the point of superposition:

$$\langle I \rangle_t = \langle I_1 \rangle_t + \langle I_2 \rangle_t + \langle I_{12} \rangle_t, \tag{10}$$

where

$$\langle I_1 \rangle_t = \langle I_2 \rangle_t = \frac{1}{2}a^2, \tag{11a}$$

$$\langle I_{12} \rangle_t = a^2 \langle \cos[\phi_1(t) - \phi_2(t) - \delta] \rangle_t. \tag{11b}$$

In deriving the averages given by Eqs. (11) we used the facts that $\langle \cos^2[\phi(t) - \bar{\omega}t] \rangle_t = 1/2$ and $\langle \cos[\phi_1(t) + \phi_2(t) - 2\bar{\omega}t + \delta] \rangle_t = 0$.

The formula (10) shows that the average intensity at the point of superposition is the sum of the averaged intensities $\langle I_1 \rangle_t$ and $\langle I_2 \rangle_t$ of each beam and of a term, $\langle I_{12} \rangle_t$, which represents the *effect of interference* between the two beams. Were the beams strictly monochromatic, the phases ϕ_1 and ϕ_2 would each be independent of time and the interference term $\langle I_{12} \rangle_t$ would differ from zero, except at certain points, for which δ takes a value that makes the cosine term vanish. It is, however, evident from the expression (11b) that an *interference term may be present even if the phases $\phi_1(t)$ and $\phi_2(t)$ are not constant; in fact, even if they fluctuate randomly*. Suppose, for example, that $\phi_1(t)$ and $\phi_2(t)$ undergo random fluctuations but that

$$\phi_2(t) - \phi_1(t) = \beta, \qquad (12)$$

where β is a constant. Then evidently the interference term $\langle I_{12} \rangle_t$ will be non-zero, except for special values of δ, just as in the idealized case of strictly monochromatic light. The simple relation (12) may be said to be an example of *statistical similarity*[1] between the vibrations at the two points. Thus we have shown that, in order *to obtain interference, light need not be monochromatic*. It is necessary only that the interfering beams possess some statistical similarity, which is the essence of coherence. This concept then acquires a precise quantitative meaning in terms of the so-called degree of coherence of light.

1.2 Temporal coherence and the coherence time

We will now introduce by means of two well-known experiments some elementary concepts relating to coherence.

Suppose first that we divide a steady-state[2] quasi-monochromatic light beam from a source S into two beams in a Michelson interferometer (Fig. 1.2) and that the two beams are brought together after a path delay $c\,\Delta t$ (c is the speed of light in vacuum) has been introduced between them. If Δt is sufficiently small, interference fringes are formed in the detection plane $\mathcal{B}$. The appearance of the fringes is said to be a manifestation of *temporal coherence* between the two beams, because the contrast of the fringes depends on the time delay Δt introduced between them. It is known from experiments that interference fringes will be observed only as long as

$$\Delta t \lesssim \frac{2\pi}{\Delta \omega}. \qquad (1)$$

---

[1] A precise meaning of the concept of statistical similarity can be given in the framework of the theory of stationary random processes. The concept is relevant not only in connection with coherence but also in connection with polarization. [H. Roychowdhury and E. Wolf, *Opt. Commun.* **248** (2005) 327–332.]

[2] By steady state we mean here that the averaged intensity

$$\frac{1}{2T} \int_{-T}^{T} I(t_0 + t)\mathrm{d}t$$

taken over an interval $2T \gg 1/\Delta\omega$ is independent of the choice of t_0. A more precise definition involves the concept of statistical stationarity which we will encounter in Section 2.1. Whilst most ordinary laboratory sources and sources found in nature (e.g. the stars) are of this kind, laser pulses do not belong to this category.

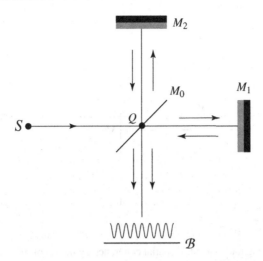

Fig. 1.2 Illustrating the concept of coherence time of light by means of the Michelson interferometer. M_1 and M_2 are mirrors. M_0 is a beam splitter (semi-transparent mirror). For the sake of simplicity, a compensating plate and a collimating lens system are not shown. To obtain interference fringes, one of the mirrors must be tilted with respect to the axis.

This result may be understood theoretically by considering the interference pattern formed by each spectral component of the light and estimating the time delay for which the individual intensity patterns get out of step, eventually canceling out. This time delay is called the *coherence time* of the light and the corresponding path delay

$$\Delta\ell = \frac{2\pi c}{\Delta\omega} = \left(\frac{\bar{\lambda}}{\Delta\lambda}\right)\bar{\lambda}, \qquad (2)$$

$\bar{\lambda}$ being the mean wavelength and $\Delta\lambda$ the effective wavelength range. This quantity is called the *coherence length* of the light.

We will illustrate the formula (2) by simple examples. Consider first thermal light, such as would be generated by incandescent matter or gas discharge with a broad spectrum. The bandwidth $\Delta\omega$ is typically of the order of $10^8 \, \mathrm{s}^{-1}$, the coherence time Δt is of the order of $10^{-8} \, \mathrm{s}$ and its coherence length $\Delta\ell \sim 2\pi \times 3 \times 10^{10} \, \mathrm{cm \, s}^{-1}/10^8 \, \mathrm{s}^{-1} \sim 19 \, \mathrm{m}$.

Let us compare this result with light generated by a well-stabilized laser, with $\Delta\omega \sim 10^4 \, \mathrm{s}^{-1}$. The coherence time of such light is of the order of $10^{-4} \, \mathrm{s}$ and its coherence length $\Delta\ell \sim 190 \, \mathrm{km}$, i.e. 10^4 times longer.

1.3 Spatial coherence and the coherence area

Another experiment which elucidates some aspects of coherence is Young's interference experiment. Suppose that quasi-monochromatic light, assumed for now to originate in a thermal source S, illuminates two pinholes in an opaque screen $\mathcal{A}$ (Fig. 1.3). For simplicity we assume that the arrangement is symmetric, with the source having the form of a square

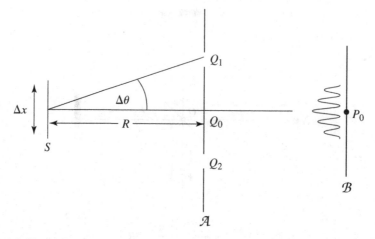

Fig. 1.3 Illustrating the concept of spatial coherence by means of Young's interference experiment with steady-state light.

of side Δx. If the pinholes at Q_1 and Q_2 are sufficiently close to each other, interference fringes will be observed on a detecting screen $\mathcal{B}$ throughout a neighborhood of the axial point P_0, provided that

$$\Delta x \, \Delta \theta < \bar{\lambda}, \tag{1}$$

where $2 \Delta \theta$ is the angle which the line $Q_1 Q_2$ subtends at the source. A rough elementary derivation of Eq. (1) may be obtained by considering the intensity distribution at the observation plane $\mathcal{B}$ to arise from the superposition of intensities of independent interference patterns formed by light from different elements of the thermal source. The relation (1) then follows on requiring that the individual interference patterns remain approximately in step.

If the plane $\mathcal{A}$ of the pinholes is at a distance R from the source plane, the fringes will be observed in the neighborhood of the axial point P_0 in the observation plane $\mathcal{B}$, provided that the pinholes are situated within an area $\Delta \mathcal{A}$ of size

$$\Delta \mathcal{A} \sim (R \, \Delta \theta)^2 = \frac{R^2}{S} \bar{\lambda}^2, \tag{2}$$

where $S = (\Delta x)^2$ is the area of the source and the relation (1) was used. This area is said to be the *coherence area* of the light in the plane $\mathcal{A}$ around the axial point Q_0. We see that it increases quadratically with the distance R from the source plane. However, the solid angle $\Delta \Omega$ which the coherence area subtends at the source is independent of the distance, being given by the formula

$$\Delta \Omega \sim \frac{\Delta \mathcal{A}}{R^2} = \frac{\bar{\lambda}^2}{S}. \tag{3}$$

Although the preceding discussion has been confined to light from a thermal source, the concept of coherence area applies more generally. However, the formula (2) applies only when the pinholes are illuminated by thermal light.

It is sometimes useful to express the coherence area in an alternative form that involves the solid angle

$$\Delta\Omega' = \frac{S}{R^2},\tag{4}$$

which the source subtends at the axial point Q_0. It follows at once from Eqs. (4) and (2) that

$$\Delta\mathcal{A} = \frac{\bar{\lambda}^2}{\Delta\Omega'}.\tag{5}$$

It is of interest to compare the coherence areas generated by various thermal sources. Let us consider first a planar thermal source of area $1\,\text{mm}^2$ emitting quasi-monochromatic thermal light of mean wavelength $\bar{\lambda} = 5,000\,\text{Å}$ (500 nm) and illuminating a plane $\mathcal{A}$ parallel to it, at a distance $R = 2\,\text{m}$. According to formula (2) the coherence area in that plane is

$$\Delta\mathcal{A} = \frac{(2\,\text{m})^2}{10^{-6}\,\text{m}^2} \times (5 \times 10^{-7}\,\text{m})^2 = 1\,\text{mm}^2.\tag{6}$$

Let us compare this result with the area of coherence of filtered sunlight illuminating the Earth's surface. Suppose that the sunlight is passed through a filter with a narrow passband around $\bar{\lambda} = 5,000\,\text{Å}$. The angular radius that the Sun's disk subtends at the surface of the Earth is approximately $\alpha \approx 0°16' \approx 0.00465$ radians. Hence, if we neglect limb darkening, the solid angle $\Delta\Omega'$ that the Sun's disk subtends at the Earth's surface is, therefore,

$$\Delta\Omega' \approx \pi\alpha^2 \approx 3.14 \times (4.65 \times 10^{-3})^2 \approx 6.79 \times 10^{-5}\,\text{sr}.\tag{7}$$

Hence, according to Eq. (5), the coherence area of sunlight on the Earth's surface is

$$\Delta\mathcal{A} = \frac{(5 \times 10^{-5}\,\text{cm})^2}{6.79 \times 10^{-5}} \sim 3.68 \times 10^{-3}\,\text{mm}^2.\tag{8}$$

Hence the linear dimension of the coherence area on the surface of the Earth of filtered sunlight is of the order of $(3.68 \times 10^{-3})^{1/2}\,\text{mm} \sim 0.061\,\text{mm}$. It is of some historical interest to mention that this value agrees with an estimate made in the 1860s by the French scientist E. Verdet, who used only a primitive notion of the concept of coherence.

Let us compare the coherence area of sunlight on the Earth's surface with that produced by a distant star. For this purpose we recall that according to Eq. (5) the coherence area is inversely proportional to the solid angle $\Delta\Omega'$ which the source subtends at the axial point Q_0 of the plane where the coherence area is to be estimated. Since the angular diameter of a star when viewed from the Earth's surface is many orders of magnitude smaller than that of the Sun, one must expect that the area of coherence of star light at the Earth's surface will be very much greater than that of the sunlight. As an example, let us consider the coherence area formed on the Earth's surface by the star Betelgeuse (α-Orionis), which was the first star whose angular diameter was determined by an interferometric technique (discussed in Section 3.3.1). It was found to have an angular diameter of $2\alpha \sim 0.047$ seconds of arc (2.3×10^{-7} radians). The solid angle which this star subtends at the Earth's surface is, therefore, $\Delta\Omega' = \pi\alpha^2 \sim 4.15 \times 10^{-14}$ sr. Hence the coherence area on the Earth's surface of the light from this star, when passed through a filter which transmits a narrow band around the wavelength $\bar{\lambda} = 5{,}000\,\text{Å}$, is, according to Eq. (5),

$$\Delta\mathcal{A} \sim \frac{(5 \times 10^{-7}\,\text{m})^2}{4.15 \times 10^{-14}\,\text{sr}} \sim 6\,\text{m}^2. \tag{9}$$

It is evident from this estimate that there is appreciable "statistical similarity" (spatial coherence) between vibrations in the filtered light from the star reaching the Earth's surface up to a separation of about $\sqrt{6\,\text{m}^2} \sim 2.45\,\text{m}$. Many stars have angular diameters which are appreciably smaller than that of Betelgeuse and, consequently, light reaching the Earth from such stars will be spatially coherent over much larger areas.

These results make it clear why stellar images formed in the focal plane of a telescope have the appearance of diffraction images formed by spatially coherent light; for our analysis shows that there exists a high degree of correlation in the star light entering the aperture of the telescope over areas which are generally much larger than the area of the aperture.

1.4 The coherence volume

Let us now consider a field that is approximately a plane, quasi-monochromatic, steady-state (i.e. statistically stationary) wave. The right-angled cylinder whose base is the coherence area in a plane normal to the direction of propagation of the light and whose height is the coherence length may be called the *coherence volume* (Fig. 1.4). It occupies a domain of space whose volume is

$$\Delta V = \Delta\ell \, \Delta A \tag{1}$$

The coherence length $\Delta\ell$ is given by Eq. (2) of Section 1.2 and the coherence area, for thermal light, is given by Eq. (2) of Section 1.3. On substituting from these equations into Eq. (1) we obtain the following expression for the coherence volume of thermal light:

$$\Delta V = \frac{R^2}{S}\left(\frac{\bar{\lambda}}{\Delta\lambda}\right)\bar{\lambda}^3 \tag{2a}$$

or, if Eq. (4) of Section 1.3 is used,

$$\Delta V = \frac{1}{\Delta\Omega'}\left(\frac{\bar{\lambda}}{\Delta\lambda}\right)\bar{\lambda}^3. \tag{2b}$$

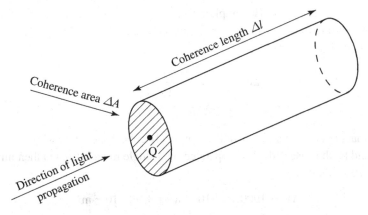

Fig. 1.4 Illustrating the concept of coherence volume ΔV in a steady-state field of a thermal, quasi-monochromatic, approximately plane wave.

Let us estimate the coherence volume for the examples we considered a moment ago in connection with the coherence area. We will assume that in each case the effective wavelength range of the filtered light $\Delta\lambda = 5 \times 10^{-4}$Å with mean wavelength $\bar{\lambda} = 5{,}000$Å. In this case the coherence length is, according to Eq. (2) of Section 1.2, about 5 m. For a planar thermal source of area $1\,\text{mm}^2$ we found [Eq. (6) of Section 1.3] that the area of coherence around the axial point Q_0 in a plane parallel to the source and at a distance $R = 2\,\text{m}$ from it has the size $\Delta A = 1\,\text{mm}^2$. Hence, according to Eq. (1) of Section 1.4 the coherence volume

$$\Delta V \sim (10^{-1}\text{cm})^2 \times (5 \times 10^2\text{cm}) = 5\,\text{cm}^3. \tag{3}$$

For filtered sunlight reaching the Earth we found [Eq. (8) of Section 1.3] that $\Delta A \sim 3.68 \times 10^{-3}\text{mm}^2$ and hence, using Eq. (1), the coherence volume of sunlight on Earth's surface is

$$\Delta V \sim 3.68 \times 10^{-3}\,\text{mm}^2 \times 5 \times 10^3\,\text{mm} \sim 18\,\text{mm}^3. \tag{4}$$

For filtered light from the star Betelgeuse, we found that $\Delta A \sim 6\,\text{m}^2$ [Eq. (9) of Section 1.3], so that

$$\Delta V \sim 6\,\text{m}^2 \times 5\,\text{m} = 30\,\text{m}^3. \tag{5}$$

The formulas (2) pertain to radiation from a thermal source. However, the concept of coherence volume applies much more generally. Consider, for example, light from a common-type laboratory helium–neon laser. Let us assume that the cross-section of the laser beam is $1\,\text{mm}^2$ and that the mean wavelength of the light is $\bar{\lambda} = 6 \times 10^{-5}\text{cm}$. Over a short time interval, of the order of a few seconds, one can easily achieve stability that ensures a narrow bandwidth $\Delta\omega \sim 10^6$ Hz, implying an effective wavelength range $\Delta\lambda \sim 1.2 \times 10^{-13}$ cm. According to Eq. (2) of Section 1.2 the coherence length during such a short time interval will be of the order of

$$\Delta\ell = 2\pi \times \frac{3 \times 10^8 \text{ m s}^{-1}}{10^6 \text{ s}^{-1}} \tag{6}$$
$$\approx 1900 \text{ m}.$$

Assuming that the laser beam is spatially completely coherent over its whole cross-section (which would be the case if the laser operated on a single mode), Eq. (1) then implies that the coherence volume is

$$\Delta V \sim 1900 \text{ m} \times 10^{-6} \text{m}^2 = 1.9 \times 10^{-3} \text{m}^3. \tag{7}$$

The concept of coherence volume has a counterpart in the quantum theory of radiation, where it represents the so-called *cell of phase space*. It is the domain in a six-dimensional phase space, formed by the three position coordinates and three momentum coordinates, throughout which photons are indistinguishable. We will briefly discuss it later [Appendix I(a)].

PROBLEMS

1.1 N quasi-monochromatic real scalar waves are superposed at a point P. The waves have constant amplitudes and the same mean frequencies. Derive an expression for the time-averaged intensity at the point P when the phases of the waves
   (1) are constant;
   (2) vary in unison, i.e. they differ from each other only by constants;
   (3) fluctuate randomly and independently of each other.

1.2 The spectral density of blackbody radiation is given by Planck's law

$$S(\nu) = \frac{8\pi h\nu^2}{c^3} \frac{1}{e^{h\nu/(k_B T)} - 1}$$

   where h is Planck's constant, k_B is the Boltzmann constant, c is the speed of light in vacuum, T is the absolute temperature and $\nu = \omega/(2\pi)$ is the circular frequency.
   Estimate the coherence length of a beam of radiation whose spectrum is given by Planck's law and plot it as a function of $k_B T$.

1.3 Betelgeuse in the constellation of Orion was the first star whose angular diameter was measured. It was found to be 0.047 seconds of arc. Determine the coherence area of the filtered light from this star on the Earth's surface, at the mean wavelength $\bar{\lambda} = 5.75 \times 10^{-5}$ cm.

2

Mathematical preliminaries

2.1 Elementary concepts of the theory of random processes

The elementary concepts that we have introduced rather heuristically in the preceding chapter can be made more precise with the help of mathematical methods developed for the analysis of random phenomena. This branch of mathematics is known as the *theory of random processes* or the *theory of stochastic processes*. In this introductory text we will not delve too deeply into the mathematical theorems and their proofs; rather we will try to get acquainted with only the basic concepts and results of the theory in order to help us to gain a deeper understanding of optical coherence and polarization effects.

Let us consider a real field variable $x(t)$. It may represent, for example, a Cartesian component of a steady-state electric field at some point in space, at time t. The exact behavior of $x(t)$ over the course of time cannot be predicted, because, as we noted in the preface, any realistic field always undergoes random fluctuations.

Suppose that $x(t)$ is measured in a series of very similar experiments and that $^1x(t)$, $^2x(t)$, $^3x(t)$, ... are the outcomes of such experiments, as indicated in Fig. 2.1. We then speak of an *ensemble of realizations* or of an ensemble of *sample functions* of the random function $x(t)$. Of course, such measurements cannot be performed at optical frequencies because of the rapidity of variation of the field, but conceptually the existence of the ensemble is clear and experiments of this kind can be performed with radiation at lower frequencies.

We have already encountered the concept of a time average [Eq. (9) of Section 1.1]. For a typical realization $^kx(t)$ of the ensemble, the time average is defined as

$$\langle\, ^kx(t)\rangle_t \;=\; \lim_{T\to\infty}\frac{1}{2T}\int_{-T}^{T}\, {}^kx(t)\mathrm{d}t. \tag{1}$$

One may define, in a similar way, the time average of a deterministic function $F(x)$, where x is a sample function $^kx(t)$ of the ensemble of x:

$$\langle F[^kx(t)]\rangle_t \;=\; \lim_{T\to\infty}\frac{1}{2T}\int_{-T}^{T} F[^kx(t)]\mathrm{d}t. \tag{2}$$

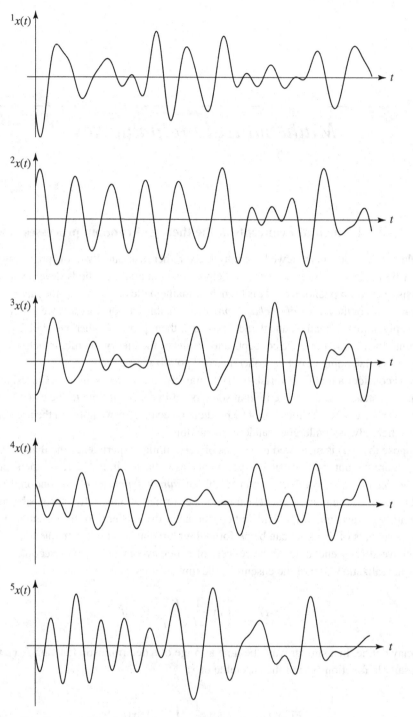

Fig. 2.1 An ensemble of realizations (sample functions) $^jx(t)$ ($j = 1, 2, \ldots$) of a random variable $x(t)$.

For example, if $F(x) = x^2$ then

$$\langle F[^k x(t)]\rangle_t = \lim_{T\to\infty} \frac{1}{2T}\int_{-T}^{T}[^k x(t)]^2\,dt. \tag{3}$$

In the theory of random processes one also defines another kind of average, called the *ensemble average* or the *expectation value*, which may be introduced as follows. Consider first the averaged sum

$$\frac{1}{N}[^1 x(t) + {}^2 x(t) + \cdots + {}^N x(t)], \tag{4}$$

and let us formally proceed to the limit as $N\to\infty$. One then writes

$$\langle x(t)\rangle_e = \lim_{N\to\infty}\frac{1}{N}\sum_{k=1}^{N} {}^k x(t), \tag{5}$$

which is one form of the ensemble average (average over an ensemble of realizations) of $x(t)$, denoted by the angular brackets with subscript e.

In a more abstract way one defines the ensemble average by means of an integral rather than by a sum. For this purpose one introduces the probability density, $p_1(x, t)$, with the following meaning. The quantity $p_1(x, t)dx$ represent the probability that the random field variable will take on a value in the range $(x, x + dx)$ at time t. The probability density p_1 can be estimated from an ensemble of realizations of x (Fig. 2.1). Suppose that among a large number N of realizations the value of x is found within the range $(x, x + dx)$ n times. Then

$$p_1(x, t) \sim \frac{n}{N}. \tag{6}$$

The ensemble average of $x(t)$, which may be regarded as a natural generalization of the definition (5), is then defined, in view of Eq. (6), as

$$\langle x(t)\rangle_e = \int x p_1(x, t)dx, \tag{7}$$

where the integration extends over the whole range of values which x can take. We stress that, although the random variable x is time-dependent, the variable x of integration under the integration sign on the right-hand side of Eq. (7) is taken to be independent of time.

The formula (7) defines the simplest kind of ensemble average, namely the mean of $x(t)$. We will also need more general averages, for example, the ensemble average of $F[x(t)]$, where $F(x)$ is again a deterministic function of x such as, for example, x^2 or $\sin x$. The ensemble average is then defined by an obvious generalization of the definition (7), namely as

$$\langle F[x(t)]\rangle_e = \int F(x)p_1(x, t)dx. \tag{8}$$

Before proceeding further with the account of the basic aspects of the theory of random processes we mention the form of the probability densities for thermal light and for laser light. For thermal light such as generated, for example, by incandescent matter or a gas discharge, the probability density of the field variable, U say, is[1]

$$p_1(U) = \frac{1}{\sqrt{2\pi\langle I \rangle}} e^{-U^2/(2\langle I \rangle)}, \tag{9}$$

where $\langle I \rangle = \langle U^2 \rangle$ is the average intensity, the average being taken over the statistical ensemble. On the other hand, the probability density p_1 of the output of a single-mode laser is well approximated by the expression

$$p_1(U) = \begin{cases} \dfrac{1}{\pi\sqrt{\langle I \rangle - U^2}} & \text{when } |U| < \sqrt{\langle I \rangle}, \\[2ex] 0 & \text{when } |U| > \sqrt{\langle I \rangle}, \end{cases} \tag{10}$$

as was first shown by L. Mandel.[2]

Plots of the two probability densities (9) and (10) are shown in Fig. 2.2. We note some basic differences between them. For thermal light, the value $U = 0$ is the most probable value, whereas for the output of a well-stabilized laser it is the least probable value, the most probable value being the square root $U = \pm\sqrt{\langle I \rangle}$ of the stabilized intensity of the laser output. Also, for thermal light there is always a finite (though possibly very small) probability that U will take on any value, whereas for the output of a single-mode laser there is zero probability that $|U| > \sqrt{\langle I \rangle}$.

Returning to the general case, the probability density $p_1(x, t)$ makes it possible to define some ensemble averages, but is by no means adequate to characterize the statistical properties of a random function. This is because $p_1(x, t)$ depends on only one time argument. It cannot, therefore, provide answers to questions concerning the statistical behavior of a random variable at several instants of time $t_1, t_2, \ldots, t_n$, and consequently about averages of

---

[1] This result is a consequence of a general theorem of probability theory, known as the *central limit theorem* (see M&W, Section 1.5.6). It asserts that, under very general conditions, the probability distribution of the sum of N independent, or weakly dependent, random variables tends, with increasing N, towards a Gaussian distribution. If we identify the random variables with the contributions to the total field arising from the different source elements (atoms), which at usual laboratory temperatures radiate independently by the process of spontaneous emission of radiation, Eq. (9) follows.

[2] L. Mandel in *Quantum Electronics*, Proc. Third International Congress, N. Bloembergen and P. Grivet eds. (Columbia University Press, New York; Dunod, Paris, 1964), pp. 101–109. See also L. Mandel and E. Wolf, *Rev. Mod. Phys.* **37** (1965), 253. The same distribution was derived from a somewhat different model by J. W. Goodman, *Statistical Optics* (J. Wiley, New York, 1985), Section 4.4.1.

The singular behavior of the probability density (10) when $|U| = \sqrt{\langle I \rangle}$ is a consequence of the rather idealized model of the output of a single-mode laser. When a more realistic model is used, which takes into account the presence of phase fluctuations caused, for example, by thermal fluctuations or vibrations of the mirrors at the ends of the laser cavity, the probability distribution will not be infinite when $|U| = \sqrt{\langle I \rangle}$ but rather it will have a sharp maximum. Plotted as a function of arg U the probability will then have the well-known doughnut shape [see, for example, O. Svelto, *Principles of Lasers*, third edition (Plenum Press, New York, 1989), Section 7.4].

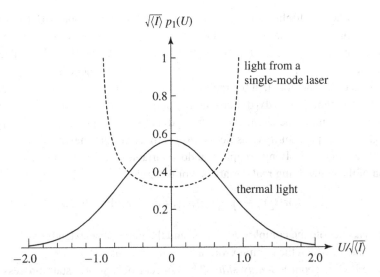

Fig. 2.2 The probability density $p_1(U)$ of the field variable U for light from a thermal source and from a single-mode laser. [Adapted from L. Mandel in *Quantum Electronics*, Proc. Third International Congress, Eds. N. Bloembergen and P. Grivet (New York, Columbia University Press and Paris, Dunod, 1964, pp. 101–109).]

products such as $x(t_1)x(t_2)$. To characterize a random process one needs a sequence of probability densities

$$p_1(x_1; t_1), \quad p_2(x_1, x_2; t_1, t_2), \quad p_3(x_1, x_2, x_3; t_1, t_2, t_3), \ldots$$

The quantity $p_2(x_1, x_2; t_1, t_2)dx_1\,dx_2$ represents the probability that the variable x will take on a value in the range $(x_1, x_1 + dx_1)$ at time t_1, and a value in the range $(x_2, x_2 + dx_2)$ at time t_2. The higher-order probability densities $p_3, p_4, \ldots$, have a similar meaning. One can then answer, for example, questions such as "what is the ensemble average of the product $x(t_1)x(t_2)$?" It is given by the formula

$$\langle x(t_1)x(t_2)\rangle_{\mathrm{e}} = \int \int x_1 x_2 p_2(x_1, x_2; t_1, t_2)dx_1\,dx_2. \tag{11}$$

We have assumed up to now that the random function is a real function of t. Later we will frequently encounter complex random functions of t, say

$$z(t) = x(t) + iy(t), \tag{12}$$

where $x(t)$ and $y(t)$ are real. The concepts that we have just been discussing may readily be generalized to such situations, as we will now indicate.

The statistical properties of a complex random process $z(t)$ are characterized by a sequence of probability densities

$$p_1(z_1; t_1), \quad p_2(z_1, z_2; t_1, t_2), \quad p_3(z_1, z_2, z_3; t_1, t_2, t_3), \ldots$$

The first one, when multiplied by $d^2z = dx_1 dy_1$, represents the probability that at time t_1 the random variable z will take on a value represented by a point located in the element $d^2z_1 = dx_1 dy_1$ around $z_1 = x_1 + iy_1$ in the complex z_1-plane (see Fig. 2.3). The second probability density, $p_2(z_1, z_2; t_1, t_2)$, when multiplied by the product $d^2z_1 d^2z_2 \equiv dx_1 dy_1 dx_2 dy_2$, represents the probability that the random variable z will take on a value represented by a point in the element $d^2z_1 = dx_1 dy_1$ around the point $z_1 = x_1 + iy_1$ at time t_1 and a value represented by a point in the element $d^2z_2 = dx_2 dy_2$ around the point $z_2 = x_2 + iy_2$ at time t_2. The higher-order probability densities $p_3, p_4, \ldots$ have similar meanings.

Ensemble averages involving complex random variables are defined by an obvious generalization of those involving real variables. For example

$$\langle z^*(t_1) z(t_2) \rangle = \int z_1^* z_2 p_2(z_1, z_2; t_1, t_2) d^2z_1 d^2z_2, \tag{13}$$

where asterisks denote the complex conjugate and the integration extends over all possible values of z_1 and z_2 which the complex variable z can take on.

We spoke earlier about a *steady state*. The concept of a steady-state process may be regarded as a non-technical term for what is known as a *stationary random process*. By this is meant a random process whose properties do not depend on the origin of time; more precisely, a random process for which all the probability densities $p_1, p_2, p_3, \ldots$ remain invariant under translation of the origin of time. Expressed mathematically, this means that

$$p_n(z_1, z_2, \ldots, z_n; t_1 + \tau, t_2 + \tau, \ldots, t_n + \tau) = p_n(z_1, z_2, \ldots, z_n; t_1, t_2, \ldots, t_n) \tag{14}$$

for all values of τ and for all positive integers n. Examples of such a situation are light vibrations at a point in the focal plane of a telescope forming the image of a star (ignoring the possibly finite life-time of the star).

For a real stationary field $U(t)$ the average intensity, $\langle I(t) \rangle$, will be independent of time, because

$$\langle I(t) \rangle_e = \langle U^2(t) \rangle_e = \int U^2 p_1(U, t) dU, \tag{15}$$

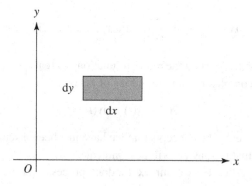

Fig. 2.3 Illustrating the meaning of the element d^2z relating to the probability $p_1(z, t) d^2z$ ($d^2z = dx\, dy$).

and p_1 does not depend on t because it has to remain unchanged under translation of the origin of time. Similar remarks apply to a complex stationary random field $V(t)$, for which the average intensity $\langle I(t) \rangle_e = \langle V^*(t)V(t) \rangle_e$.

We are now familiar with two kinds of averages, the time average and the ensemble average. Fortunately it turns out that, when one is dealing with stationary fields, these two averages usually have the same value. This is the essence of so-called *ergodicity*, a concept which we will now briefly discuss.

2.2 Ergodicity

In a broad sense a statistically stationary process is said to be *ergodic* if the time average, taken over the interval $-\infty < t < \infty$, of any deterministic function $F(\xi)$ of a typical realization $\xi = {}^k x(t)$ of the random process is equal to the corresponding ensemble average, i.e. if

$$\langle F[{}^k x(t)] \rangle_t = \langle F[x(t)] \rangle_e; \tag{1}$$

and, more generally, if $F(y_1, y_2, \ldots, y_n)$ is a deterministic function of several variables y_1, $y_2, \ldots, y_n$, then, for an ergodic ensemble

$$\langle F[{}^k x(t_1), {}^k x(t_2), \ldots, {}^k x(t_n)] \rangle_t = \langle F[x(t_1), x(t_2), \ldots, x(t_n)] \rangle_e. \tag{2}$$

Such equalities may seem to be somewhat surprising at first sight since, for example, in Eq. (1) the expression on the left-hand side seems to depend on the particular choice ${}^k x(t)$ of a realization of the random process and is independent of time, whereas the expression on the right-hand side does not depend on any particular realization and seems to depend on time. The following example may perhaps indicate why the equality of the two types of averages might be expected.

Consider a particular realization ${}^k x(t)$ of a stationary random process, such as that shown in Fig. 2.4(a). Let us imagine that we divide it into long segments, which we label as ①, ②, ③, . . ., each of duration $2T$ and let us place them under each other, as in Fig. 2.4(b). For an ergodic ensemble we may expect that all the individual segments will be representative of the ensemble of $x(t)$ of which ${}^k x(t)$ is a member. Since the ensemble shown in Fig. 2.4(b) was derived from the single realization ${}^k x(t)$ of the original ensemble, one might expect that the statistical information which can be deduced from the single realization, (a), and from the ensemble of realizations, (b), will be the same.

We will assume from now on that we are dealing with ensembles that are stationary and ergodic. Consequently, it will not be necessary to distinguish between time averages and ensemble averages and we will, therefore, omit from now on the subscripts t and e on the angular brackets that denote expectation values.

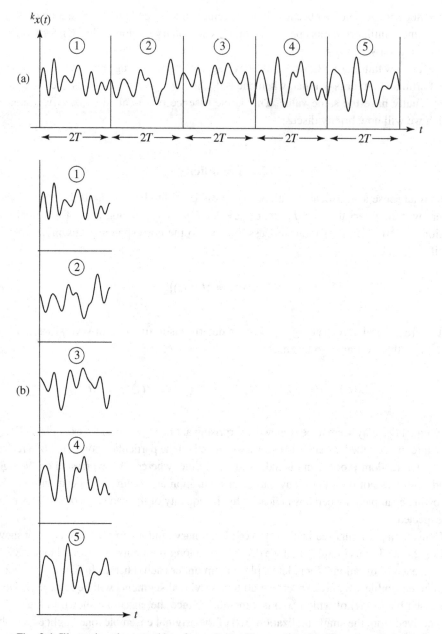

Fig. 2.4 Illustrating the meaning of ergodicity. The statistical information about the ensemble of a stationary random process $x(t)$ is contained in a single sample function $^kx(t)$ of the process. Dividing such a sample function into many parts, ①, ②, ③, . . ., [(a)] provides a valid ensemble [(b)] of realizations (sample functions) of $x(t)$.

2.3 Complex representation of a real signal and the envelope of a narrow-band signal

In the usual treatments, when one deals with a real monochromatic wavefield, it is often convenient to associate with it a complex monochromatic wavefield (see, for example, B&W, Section 1.4.3). The use of the associated complex field simplifies calculations, especially if they involve averages of quantities that are quadratic in the field variable, e.g. the intensity. There is a similar procedure, which we will now describe, which is useful when one is dealing with real random fields such as, for example, the wavefield generated by a thermal source.

Let $U(t)$ be a fluctuating real quantity, which might represent, for example, a Cartesian component of the electric field at some point in space. Let us express it in the form of a Fourier integral,

$$U(t) = \int_{-\infty}^{\infty} u(\omega)e^{-i\omega t}d\omega. \tag{1}$$

Since $U(t)$ is real, it is equal to its complex conjugate and, using this fact, it readily follows that

$$u(-\omega) = u^*(\omega), \tag{2}$$

where the asterisk denotes the complex conjugate. This result implies that the negative frequency components contain no information that is not already contained in the positive ones. For this reason we may omit the negative frequency components, i.e. we may use in place of the function $U(t)$ the function

$$V(t) = \int_0^{\infty} u(\omega)e^{-i\omega t}d\omega. \tag{3}$$

$V(t)$ is known as the *complex analytic signal* associated with the real signal $U(t)$. The name derives from the fact that the function $V(t)$, when considered as a function of *complex t*, may be shown to be analytic in the lower part of the complex t plane (M&W, Section 3.1). Moreover, one can also show that [M&W, Eq. (2.1-57)]

$$V(t) = \frac{1}{2}[V^{(r)}(t) + iV^{(i)}(t)], \tag{4}$$

where the functions $V^{(r)}(t) = U(t)$ and $V^{(i)}(t)$ form a so-called Hilbert-transform pair (sometimes referred to as a conjugate pair); and, moreover, if $V^{(r)}(t)$ is a stationary random variable, then

$$\langle [V^{(r)}(t)]^2 \rangle = \frac{1}{2}\langle V^*(t)V(t)\rangle. \tag{5}$$

Since the expectation value often represents the average intensity, the formula (5) shows that the expectation value $\langle V^*(t)V(t)\rangle$ may also be taken to be a measure of the average intensity.

Analytic signals are particularly useful in representing the envelope of narrow-band fields. To see this let us first consider a strictly monochromatic real signal $x(t)$ of frequency ω_0, i.e.

$$x(t) = a \cos(\psi - \omega_0 t), \tag{6}$$

where a and ψ are the amplitude and phase factors of the signal, both being constants. We may rewrite Eq. (6) in the form

$$x(t) = \frac{1}{2} \xi_0 e^{-i\omega_0 t} + \frac{1}{2} \xi_0^* e^{i\omega_0 t}, \tag{7}$$

where

$$\xi_0 = a e^{i\psi}. \tag{8}$$

The formula (7) implies that the Fourier spectrum of $x(t)$ consists of two delta functions, centered at frequencies ω_0 and $-\omega_0$ [see Fig. 2.5(a)].

Next let us suppose that $x(t)$ is not strictly monochromatic but rather that it has a narrow bandwidth $\Delta\omega$ which is small relative to its mean frequency $\bar{\omega}$, i.e. that

$$\frac{\Delta\omega}{\bar{\omega}} \ll 1. \tag{9}$$

Such a narrow-band signal is said to be *quasi-monochromatic*. A typical Fourier spectrum of such a signal is shown in Fig. 2.5(b). One often represents it in a form that is a generalization of that given by Eq. (6), namely as

$$x(t) = a(t)\cos[\psi(t) - \bar{\omega}t]. \tag{10}$$

The amplitude and the phase of such a signal are no longer constant as they are for a monochromatic signal, rather they vary with time. However, as is intuitively perhaps obvious and can be verified by elementary Fourier analysis, their variations are very slow compared with the variation due to the term $\bar{\omega}t$, remaining essentially constant over an interval that is small compared with the reciprocal bandwidth $1/\Delta\omega$. It should be noted, however, that the representation (10) is somewhat ambiguous, because it associates with a real function $x(t)$ two real functions, $a(t)$ and $\psi(t)$, and such an association is not unique. The analytic signal representation makes it possible to avoid the non-uniqueness. One defines the complex envelope by associating with $x(t)$ the analytic signal [*cf.* Eq. (4)]

$$z(t) = \frac{1}{2}[x(t) + iy(t)], \tag{11}$$

where $y(t)$ is the conjugate (Hilbert transform) of $x(t)$. We can express $y(t)$ in the form

$$y(t) = a(t)\sin[\psi(t) - \bar{\omega}t]. \tag{12}$$

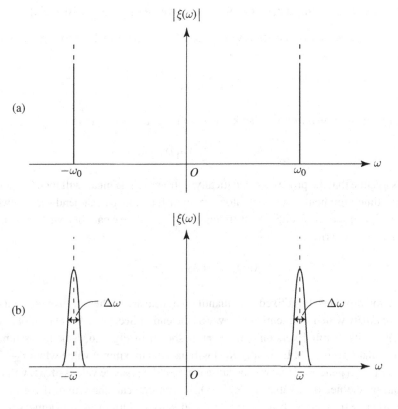

Fig. 2.5 The Fourier spectrum of a monochromatic signal of frequency ω_0 is shown in
(a) and that of a real quasi-monochromatic signal of mean frequency $\bar{\omega}$ and bandwidth
$\Delta\omega \ll \bar{\omega}$ is shown in (b).

We now have two equations, namely (12) and (10), from which $a(t)$ and $\psi(t)$ may be
uniquely determined (apart from a trivial ambiguity of an additive phase factor $2m\pi$, where
m is an integer).

The function

$$2z(t) = [x(t) + iy(t)]$$
$$= a(t)e^{i[\psi(t) - \bar{\omega}t]} \tag{13}$$

may be identified with a *complex envelope of the narrow-band signal* $x(t)$.

We have so far regarded $x(t)$ as a deterministic function. Suppose instead that $x(t)$ is a stationary quasi-monochromatic random function whose spectral density has an effective
bandwidth $\Delta\omega$ which is much smaller than the mean frequency $\bar{\omega}$. Such a random function
can be represented by an ensemble of quasi-monochromatic signals. It is clear that its statistical properties are entirely reflected in the statistical behavior of the amplitude function
$a(t)$ and of the phase function $\psi(t)$.

2.4 The autocorrelation and the cross-correlation functions

The two most important expectation values associated with a real random process $x(t)$ are its mean

$$m(t) \equiv \langle x(t) \rangle \tag{1}$$

and its *autocorrelation function* (also known as the covariance function)[1]

$$R(t_1, t_2) \equiv \langle x(t_1)x(t_2) \rangle. \tag{2}$$

Let us assume that the process is statistically stationary. The mean will then be independent of the time argument t and the autocorrelation function will depend on the two time arguments t_1 and t_2 only through the difference $\tau = t_2 - t_1$. We can then write $R(\tau)$ in place of $R(t_1, t_2)$, i.e. we write

$$R(\tau) = \langle x(t)x(t + \tau) \rangle. \tag{3}$$

This function provides, for a fixed τ, a quantitative measure of the intuitive concept of *statistical similarity* which we mentioned towards the end of Section 2.1. Let us plot a realization $^k x(t)$ and its "shifted version" $^k x(t + \tau)$ as shown in Fig. 2.6, it being assumed, for simplicity, that $\langle ^k x(t) \rangle_t = 0$. Evidently $R(\tau)$ will have its maximum value when $\tau = 0$. As τ increases $R(\tau)$, in general, decreases because some of the positive values of $x(t)$ will be cancelled out by the negative values of $x(t + \tau)$. In fact one can show that, if the process is ergodic, $R(\tau) \to 0$ as $\tau \to \infty$. It is clear that the effective width of $R(\tau)$ is a measure of the time over which there exist correlations (some statistical similarity) between $x(t)$ and $x(t + \tau)$. The effective width of $R(\tau)$ is evidently a more precise measure of the concept of coherence time which we introduced earlier rather heuristically by an order-of-magnitude relation [Eq. (1) of Section 1.2].

The autocorrelation function $R(\tau)$ has a number of useful properties, of which some follow immediately from its definition and others can be readily derived. The most important ones are

$$R(0) \geq 0, \tag{4a}$$

$$R(-\tau) = R(\tau), \tag{4b}$$

$$|R(\tau)| \leq R(0). \tag{4c}$$

---

[1] If the mean $m(t)$ of $x(t)$ is not zero, one often uses instead of the autocorrelation function $R(t_1, t_2)$ the *centered* autocorrelation function

$$\tilde{R}(t_1, t_2) \equiv \langle [x(t_1) - m(t_1)][x(t_2) - m(t_2)] \rangle.$$

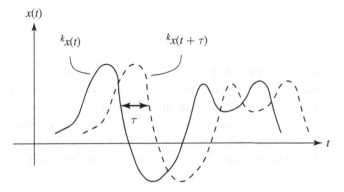

Fig. 2.6 Illustrating the significance of the autocorrelation function of a real random process $x(t)$ of zero mean, as a measure of statistical similarity. As the separation τ between the sample functions at two different instants of time t and $t + \tau$ gradually increases, ${}^k x(t)$ and ${}^k x(t + \tau)$ will acquire quite different values for each t, and the positive and the negative contributions will tend to cancel out in their product.

There is a theorem which asserts that the Fourier transform of the autocorrelation function is necessarily non-negative. This theorem, which is relevant to the definition of the spectrum of a stationary random process (see Section 2.5) is known as *Bochner's theorem* (M&W, p. 18).

When the random variable is complex, we write $z(t)$ instead of $x(t)$ and we define its autocorrelation function in a somewhat similar way to how it is defined for a real process $x(t)$. If the complex process is stationary and of zero mean, we have in place of Eq. (3)

$$R(\tau) \equiv \langle z^*(t)z(t + \tau) \rangle, \tag{5}$$

where the asterisk denotes the complex conjugate. The inequalities (4a) and (4c) still hold. Instead of the formula (4b) one has

$$R(-\tau) = R^*(\tau). \tag{6}$$

Later we will also need a generalization of the autocorrelation function for situations involving two random processes, say $z_1(t)$ and $z_2(t)$. These processes may represent, for example, the field variable at two points P_1 and P_2 in space. Assuming that the processes are also jointly stationary, i.e. that the joint probability of $z_1(t)$ and $z_2(t)$ is invariant with respect to translation of the origin of time, the measure of their correlation is the so-called *cross-correlation function*

$$R_{12}(\tau) = \langle z_1^*(t)z_2(t + \tau) \rangle. \tag{7}$$

It has the following properties:

$$|R_{12}(\tau)| \leq \sqrt{R_{11}(0)R_{22}(0)} \tag{8a}$$

and

$$R_{12}(-\tau) = R_{21}^*(\tau). \tag{8b}$$

Among the various parameters which characterize a random process, the mean $m(t)$ and the autocorrelation function $R(t_1, t_2)$ are particularly useful. When the mean is independent of time and the autocorrelation function depends only on the time difference $t_2 - t_1$ one says that the process is *stationary in the wide sense*. Obviously a process which is (strictly) stationary (i.e. for which all its probability densities remain invariant under the translation of origin of time) is also stationary in the wide sense, but the converse is, of course, not necessarily true.

2.4.1 The autocorrelation function of a finite sum of periodic components with random amplitudes

We will now examine the behavior of the autocorrelation function of a class of stationary random processes which has a bearing on the concept of the spectrum of a stationary random process, which we will discuss in Section 2.5.

Consider a random process represented by the ensemble

$$z(t) = \{^k z(t)\} \quad (k = 1, 2, 3, \ldots, M), \tag{9}$$

where each realization $^k z(t)$ is a sum of periodic components with frequencies $\omega_1, \omega_2, \ldots, \omega_M$, i.e.

$$^k z(t) = {}^k\zeta_1 e^{-i\omega_1 t} + {}^k\zeta_2 e^{-i\omega_2 t} + \cdots + {}^k\zeta_M e^{-i\omega_M t}, \tag{10}$$

the $^k\zeta_m$ ($m = 1, 2, 3, \ldots$) being (generally complex) random variables. We will assume that for each m the $^k\zeta_m$ is of zero mean, i.e. that

$$\langle {}^k\zeta_m \rangle = 0 \quad (m = 1, 2, \ldots, M), \tag{11}$$

and that the process is *stationary*, at least in the wide sense.

The autocorrelation function of this random process is

$$R(\tau) \equiv \langle z^*(t)z(t + \tau)\rangle$$

$$= \left\langle \left(\sum_{m=1}^{M} \zeta_m^* e^{i\omega_m t} \right) \left(\sum_{n=1}^{M} \zeta_n e^{-i\omega_n(t+\tau)} \right) \right\rangle$$

$$= \sum_{m=1}^{M} \sum_{n=1}^{N} \langle \zeta_m^* \zeta_n \rangle e^{-i(\omega_n - \omega_m)t} e^{-i\omega_n \tau}. \tag{12}$$

Since the process is stationary, the right-hand side must be independent of t and this can evidently be so only if

$$\langle \zeta_m^* \zeta_n \rangle = 0 \quad \text{when } n \neq m, \tag{13}$$

i.e. *the periodic terms of different frequencies must be uncorrelated*. The formula (12) then simplifies to

$$R(\tau) = \sum_{m=1}^{M} \langle \zeta_m^* \zeta_m \rangle e^{-i\omega_m \tau}. \tag{14}$$

We see that the autocorrelation function $R(\tau)$ is now the sum of periodic terms at the same frequencies $\omega_1, \omega_2, \ldots, \omega_M$ as are present in the sample functions (10) of the process and that the spectral component of each frequency ω_m is (apart from a proportionality constant depending on the choice of units) proportional to the average "energy" (or the average "power") associated with each periodic component of the process. We note that the autocorrelation function provides no information about phases of the periodic components of the sample functions. This process is evidently not ergodic because, as we mentioned earlier, for an ergodic process $R(\tau) \rightarrow 0$ as $\tau \rightarrow \infty$, a condition which is not satisfied by the expression (14).

2.5 The spectral density and the Wiener–Khintchine theorem

The Fourier spectrum, which provides information about the strength of the periodic components of various frequencies into which a function can be decomposed, is an important concept in much of physics and engineering. It can be applied to many deterministic functions encountered in practice.

The situation is more complicated with functions that are sample functions of a stationary random process. Such functions do not have a Fourier representation, because they are defined in the time interval $-\infty < t < \infty$ and do not approach zero as $t \rightarrow \infty$ and $t \rightarrow -\infty$. Attempts to decompose such functions into harmonic components, i.e. into periodic components, have a long, rich and fascinating history, which cannot be discussed here. We will only show how their formal decomposition may be used to introduce an important concept, namely the concept of the spectrum of a (wide-sense) stationary random process. We will ignore the mathematical subtleties encountered in a rigorous definition of the spectrum of such a process.[1]

Consider a wide-sense-stationary complex random process $z(t)$ of zero mean, i.e. such that

$$\langle z(t) \rangle = 0. \tag{1}$$

Let us formally represent a typical realization as a Fourier integral

$$^k z(t) = \int_{-\infty}^{\infty} {}^k \zeta(\omega) e^{-i\omega t} d\omega. \tag{2}$$

---

[1] A rigorous treatment requires the use of *distribution theory*, also called *generalized function theory*. An excellent account of this subject is given in H. M. Nussenzveig, *Causality and Dispersion Relations* (Academic Press, New York, 1972), Appendix A, pp. 362–390.

Then, on inverting Eq. (2), we have

$$^k\zeta(\omega) = \frac{1}{2\pi} \int_{-\infty}^{\infty} {}^k z(t) e^{i\omega t} dt \tag{3}$$

and, consequently, with any two frequencies ω and ω',

$$^k\xi^*(\omega){}^k\xi(\omega') = \frac{1}{(2\pi)^2} \int_{-\infty}^{\infty} \int_{-\infty}^{\infty} {}^k z^*(t){}^k z(t') e^{-i\omega t} e^{i\omega' t'} dt\, dt'. \tag{4}$$

Let us set $t' = t + \tau$, take the ensemble average of both sides and formally interchange ensemble averaging and integration. We then obtain the formula

$$\langle \xi^*(\omega)\xi(\omega') \rangle = \frac{1}{(2\pi)^2} \int_{-\infty}^{\infty} \int_{-\infty}^{\infty} R(\tau) e^{i(\omega'-\omega)t} e^{i\omega' \tau} dt\, d\tau, \tag{5}$$

where

$$R(\tau) = \langle z^*(t)z(t + \tau) \rangle \tag{6}$$

is the autocorrelation function of the random process encountered in Section 2.4. The integration over t on the right-hand side of Eq. (5) can be carried out at once, being proportional to a Dirac delta function:

$$\frac{1}{2\pi} \int_{-\infty}^{\infty} e^{i(\omega'-\omega)t} dt = \delta(\omega' - \omega). \tag{7}$$

Equation (5) therefore implies that

$$\langle \zeta^*(\omega)\zeta(\omega') \rangle = S(\omega)\delta(\omega - \omega'), \tag{8}$$

where

$$S(\omega) = \frac{1}{2\pi} \int_{-\infty}^{\infty} R(\tau) e^{i\omega\tau} d\tau. \tag{9a}$$

On taking the Fourier inverse of Eq. (9a) we have

$$R(\tau) = \int_{-\infty}^{\infty} S(\omega) e^{-i\omega\tau} d\omega. \tag{9b}$$

The formulas (8) and (9) imply two things. The first shows that for a wide-sense-stationary random process the (generalized) spectral components of different frequencies are not correlated; and that the "strength" of the "self-correlation" (for $\omega' = \omega$) is, according to Eq. (9a), equal to the Fourier transform $S(\omega)$ of the autocorrelation function $R(\tau)$ of the random process. $S(\omega)$, which is formally *defined* by Eq. (8), is known as the *spectral density* (also known as the *spectrum* or *power spectrum*, or the *Wiener spectrum*) of the (stationary) random process $z(t)$.

In order that $S(\omega)$ agrees with the intuitive notion of a spectrum, it must, of course, be non-negative. That this is so is fairly clear from the definition (8) and follows rigorously from Bochner's theorem that was mentioned earlier (Section 2.4), applied to Eq. (9a).

The results expressed by Eqs. (8) and (9) are natural generalizations of those encountered in Section 2.4, in connection with a stationary random process consisting of a finite number of periodic components of different frequencies.

The Dirac delta function on the right-hand side of Eq. (8) can be removed by integrating both sides of that equation with respect to ω' over a small range $(\omega - \frac{1}{2}\Delta\omega \leq \omega' \leq \omega + \frac{1}{2}\Delta\omega)$ and then letting $\Delta\omega \to 0$. This gives

$$ S(\omega) = \underset{\Delta\omega \to 0}{\text{Lim}} \int_{\omega - \frac{1}{2}\Delta\omega}^{\omega + \frac{1}{2}\Delta\omega} \langle \zeta^*(\omega)\zeta(\omega')\rangle d\omega'. \tag{10} $$

The integration over the small frequency range expresses a kind of smoothing. In fact smoothing is quite essential if one wishes to estimate the spectrum from a single realization of a stationary random process (see Fig. 2.7).[1]

The formulas (9) are basic relations of the theory of stationary random processes, expressing the so-called *Wiener–Khintchine theorem*. In words, this theorem expresses the fact that *the spectrum $S(\omega)$ of a wide-sense-stationary random process of zero mean and its autocorrelation function $R(\tau)$ form a Fourier-transform pair.*

For later purposes we will need a generalization of the concept of the spectral density and of the Wiener–Khintchine theorem from a single random process $z(t)$ to a pair of random processes $z_1(t)$ and $z_2(t)$ which are jointly stationary, at least in the wide sense. In optics $z_1(t)$ and $z_2(t)$ often represent the fluctuating optical field (represented by complex

---

[1] Alternatively smoothing can be provided by ensemble averaging. More specifically one can show [see, for example, S. Goldman, *Information Theory* (Prentice Hall, New York, 1955), p. 244] that

$$ S(\omega) = \underset{T \to \infty}{\text{Lim}} \left\langle \frac{|\zeta(\omega, T)|^2}{2T} \right\rangle, \tag{10a} $$

where

$$ \zeta(\omega, T) = \frac{1}{2\pi} \int_{-\infty}^{\infty} z_T(t)e^{i\omega t}dt \tag{10b} $$

is the Fourier transform of the truncated process

$$ z_T(t) = \begin{cases} z(t) & \text{when } |t| \leq T, \\ 0 & \text{when } |t| > T. \end{cases} $$

The ensemble average and the limiting process ($T \to \infty$) have to be taken in the order indicated in Eq. (10a). In fact the limit as $T \to \infty$ of the expression in the angular brackets does not, in general, exist.

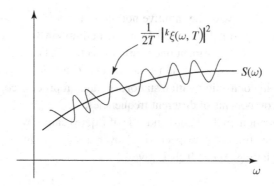

Fig. 2.7 Schematic illustration of the spectrum $S(\omega)$ of a random process, stationary at least in the wide sense, and the spectrum

$$^{k}\zeta(\omega, T) = \frac{1}{2\pi} \int_{-T}^{T} {}^{k}z(t)e^{i\omega t}\, dt$$

obtained from the Fourier transform of a single realization $^{k}z(t)$ of the process, truncated over an arbitrary interval $-T \leq t < T$. In general, the true spectrum $S(\omega)$ cannot be determined from a single realization of the process without smoothing. Taking the ensemble average, as indicated in Eq. (10a) of Section 2.5, provides one way of smoothing.

analytic signals) at two points in space. By a strictly similar analysis to that which led to the derivation of formulas (8) and (9) one finds that

$$\langle \zeta_1^*(\omega)\zeta_2(\omega') \rangle = W_{12}(\omega)\delta(\omega - \omega'),\tag{11}$$

where

$$W_{12}(\omega) = \frac{1}{2\pi} \int_{-\infty}^{\infty} R_{12}(\tau)e^{i\omega\tau}d\tau,\tag{12a}$$

and, taking the inverse,

$$R_{12}(\tau) = \int_{-\infty}^{\infty} W_{12}(\omega)e^{-i\omega\tau}d\omega.\tag{12b}$$

In Eq. (11), $\zeta_1(\omega)$ and $\zeta_2(\omega)$ are the generalized Fourier transforms of $z_1(t)$ and $z_2(t)$, respectively. The function $W_{12}(\omega)$, defined by (12a), is known as the *cross-spectral density* of $z_1(t)$ and $z_2(t)$ and the function

$$R_{12}(\tau) = \langle z_1^*(t)z_2(t + \tau) \rangle\tag{13}$$

is the cross-correlation function of the two random processes, whose main properties we noted in Section 2.4.

We will refer to the pair of formulas (12a) and (12b) as the *generalized Wiener–Khintchine theorem.*

PROBLEMS

2.1 Determine the analytic signal associated with the real signal

$$x(t) = \begin{cases} 1 & \text{when } -1 \leq t \leq 1, \\ 0 & \text{otherwise.} \end{cases}$$

2.2 $x(t)$ is a real signal, bandlimited to the range $|\omega_1| \leq |\omega| \leq |\omega_1| + \Delta|$. Show that the square $a^2 = 4|z(t)|^2$ of its real envelope is bandlimited to the range $-\Delta \leq \omega \leq \Delta$.

2.3 Consider a real random process

$$x(t) = a \sin(\omega t + \theta),$$

where a and ω are constants and θ is a random variable. Find necessary and sufficient conditions on θ to ensure that the process $x(t)$ is stationary in the wide sense. Give also a specific example of such a random variable θ.

2.4 $x(t)$ is a wide-sense-stationary real random process of zero mean. Show that the complex process $z(t)$, obtained from $x(t)$ by replacing each sample function by the associated analytic signal, is also stationary in the wide sense and is of zero mean.

2.5 A random process $x(t) = A$, where A is a random variable, which is uniformly distributed on the interval $(-1, 1)$.
(a) Sketch some sample function of the process.
(b) Determine the autocorrelation function of $x(t)$ defined
(i) by a time average; and
(ii) by an ensemble average.
(c) Is $x(t)$ wide-sense stationary? Is it strictly stationary?
(d) Is $x(t)$ an ergodic random process?
Your answers to (c) and (d) should provide justifications.

2.6 $x(t)$ is a real wide-sense stationary random process and $y(t)$ is a process derived from $x(t)$ by linear filtering, i.e. by a relation of the form

$$y(t) = \int_{-\infty}^{\infty} K(t - t')x(t')dt'.$$

Show that $y(t)$ is also a wide-sense-stationary random process and derive relations between (i) the autocorrelation functions of the two processes and (ii) the power spectra of the two processes.

2.7 Determine the autocorrelation function of the random process

$$x(t) = a_1 \cos(\omega_1 t - \phi_1) + a_2 \cos(\omega_2 t - \phi_2),$$

where a_1, a_2, ω_1 and ω_2 are known constants and ϕ_1 and ϕ_2 are mutually independent random variables, each of which is uniformly distributed on the interval $(0, 2\pi)$. Determine also the power spectrum of the process.

2.8 $z(t)$ is a complex stationary Gaussian random process of zero mean. Let

$$z_T(t) = \begin{cases} z(t) & \text{when } |t| \le T \\ 0 & \text{when } |t| > T \end{cases}$$

and let $\xi(\omega, T)$ be the Fourier inverse of $z_T(t)$. Further let

$$S_T(\omega) = \frac{\xi^*(\omega, T)\xi(\omega, T)}{2T}$$

and

$$S(\omega) = \lim_{T \to \infty} \langle S_T(\omega) \rangle.$$

Show that, at any frequency ω for which $S(\omega) \ne 0$, the variance of $S_T(\omega)$ does not tend to zero at $T \to \infty$. What is the implication of this result for the problem of determining the spectral density function of the process from one of its sample functions?

2.9 The autocorrelation function of a real stationary random process $x(t)$ is

$$R(\tau) = \begin{cases} 1 - |\tau|/T & \text{when } |t| < T, \\ 0 & \text{otherwise.} \end{cases}$$

Determine the power spectrum of the process.

2.10 Consider two random processes

$$x(t) = u\cos(\omega t) + v\sin(\omega t), \qquad y(t) = -u\sin(\omega t) + v\cos(\omega t),$$

where u and v are uncorrelated random variables with zero mean and with the same variance.

(i) Find the cross-correlation function of the two processes.
(ii) Determine whether the two processes are jointly stationary in the wide sense.
(iii) Determine whether the two processes are ergodic with respect to their cross-correlation function.

3

Second-order coherence phenomena
in the space–time domain

3.1 Interference law for stationary optical fields. The mutual coherence function and the complex degree of coherence

In Section 1.1 we pointed out that light need not be monochromatic in order to produce interference between two light beams; it is necessary only that the light vibrations in the two beams are "statistically similar" to each other. We will now introduce a quantitative measure of statistical similarity, with the help of Young's interference experiment.

Suppose that light, assumed to be stationary, at least in the wide sense, illuminates two pinholes Q_1 and Q_2 in an opaque screen $\mathcal{A}$ (Fig. 3.1). Let $V(Q_1, t)$ and $V(Q_2, t)$ represent the light vibrations at the pinholes at time t. For simplicity we consider $V(Q, t)$ to be a (complex) scalar. Generalization to vector fields will be considered in Chapter 7. We will examine the distribution of the average intensity in the neighborhood of some point P on a screen $\mathcal{B}$ behind the screen $\mathcal{A}$ containing the pinholes.

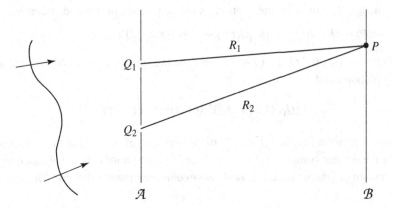

Fig. 3.1 Notation relating to Young's interference experiment with quasi-monochromatic light.

Let R_1 and R_2 denote the distances Q_1P and Q_2P respectively. Since the light from the pinholes Q_1 and Q_2 takes the times

$$t_1 = \frac{R_1}{c}, \qquad t_2 = \frac{R_2}{c} \tag{1}$$

to reach the point P, $V(P, t)$ is given by the expression

$$V(P, t) = K_1 V(Q_1, t - t_1) + K_2 V(Q_2, t - t_2), \tag{2}$$

where the factors K_1 and K_2 take into account diffraction at the pinholes and c in Eq. (1) denotes the speed of light in vacuum. It follows from the Huygens–Fresnel principle (B&W, Section 8.2) that, if the angles of incidence and diffraction at the pinholes are sufficiently small, then[1]

$$K_j \approx -\frac{i}{\bar{\lambda} R_j} d\mathcal{A}_j \quad (j = 1, 2), \tag{3}$$

where $d\mathcal{A}_1$ and $d\mathcal{A}_2$ are the areas of the two pinholes and $\bar{\lambda}$ is the mean wavelength of the incident light.

Because the fluctuations of the light are very rapid, one cannot observe the time behavior of $V(P, t)$, but one can measure the expectation value $I(P) \equiv \langle I(P, t) \rangle = \langle V^*(P, t) V(P, t) \rangle$ of the intensity, which is independent of time, because of the assumed stationarity. Using Eq. (2), it follows that

$$
\begin{aligned}
I(P) = {} & |K_1|^2 \langle V^*(Q_1, t - t_1) V(Q_1, t - t_1) \rangle \\
& + |K_2|^2 \langle V^*(Q_2, t - t_2) V(Q_2, t - t_2) \rangle \\
& + 2\,\mathcal{R}e\,\{K_1^* K_2 \langle V^*(Q_1, t - t_1) V(Q_2, t - t_2) \rangle\},
\end{aligned}
\tag{4}
$$

$\mathcal{R}e$ denoting the real part. Using the assumption of stationarity, and also the fact that K_1 and K_2 are purely imaginary numbers, the expression (4) may be simplified and becomes

$$I(P) = |K_1|^2 I(Q_1) + |K_2|^2 I(Q_2) + 2\,\mathcal{R}e\,\{|K_1||K_2| \Gamma(Q_1, Q_2, t_1 - t_2)\}, \tag{5}$$

where $I(Q_j) = \langle V^*(Q_j, t) V(Q_j, t) \rangle$ $(j = 1, 2)$ are the averaged intensities of the light at each of the two pinholes and

$$\Gamma(Q_1, Q_2, \tau) = \langle V^*(Q_1, t) V(Q_2, t + \tau) \rangle \tag{6}$$

is the cross-correlation function [Eq. (13) of Section 2.5] of the field at the pinholes at Q_1 and Q_2. In the present context $\Gamma(Q_1, Q_2, \tau)$ is called the *mutual coherence function* and is the basic quantity of the so-called *second-order coherence theory*, the term "second-order"

---

[1] It is convenient in the present analysis to define the K factors somewhat differently from how they were defined in B&W. We include here the terms $d\mathcal{A}_j/R_j$ in their definition, thus making them dimensionless.

indicating that Γ is a correlation function involving a product of the field variable at two points. Later, in Chapter 7, we will encounter correlation functions involving a four-term product, which are the basic mathematical tools of the so-called *fourth-order* coherence theory.[1]

The formula (5) may be rewritten in a physically more significant form if we note that the first two terms on the right-hand side of Eq. (5) represent the average intensities of the light that would be observed at the point P if only one of the pinholes were open. For example, if just the pinhole at Q_1 were open, one would have $K_2 = 0$ and Eq. (5) then implies that

$$I(P) \equiv I^{(1)}(P) = |K_1|^2 I(Q_1). \tag{7a}$$

Similarly, if just the pinhole at Q_2 were open, then

$$I(P) \equiv I^{(2)}(P) = |K_2|^2 I(Q_2). \tag{7b}$$

The formula (5) may, therefore, be rewritten as

$$I(P) = I^{(1)}(P) + I^{(2)}(P) + 2\,\mathcal{Re}\left[\sqrt{I^{(1)}(P)}\sqrt{I^{(2)}(P)}\gamma(Q_1, Q_2, t_1 - t_2)\right], \tag{8}$$

where

$$\gamma(Q_1, Q_2, \tau) = \frac{\Gamma(Q_1, Q_2, \tau)}{\sqrt{I(Q_1)}\ \sqrt{I(Q_2)}} \tag{9}$$

$$= \frac{\Gamma(Q_1, Q_2, \tau)}{\sqrt{\Gamma(Q_1, Q_1, 0)}\sqrt{\Gamma(Q_2, Q_2, 0)}}. \tag{10}$$

The formula (8) is one form of the so-called *interference law for stationary optical fields*. It shows that, in order to determine the (average) intensity at a point P in the observation plane $\mathcal{B}$, one must know not only the average intensities of the two beams at P but also the real part of the correlation coefficient $\gamma(Q_1, Q_2, \tau)$, called the *complex degree of coherence*, of the light at the pinholes at Q_1 and Q_2. In view of the significance of the cross-correlation function which we discussed earlier, the complex degree of coherence is evidently a precise measure of the *statistical similarity* of the light vibrations at the points Q_1 and Q_2.

It follows from the definition (10) and the property of the cross-correlation function expressed by Eq. (8a) of Section 2.4 that, for all values of its arguments, γ is bounded by zero and by unity in absolute value, i.e. that

$$0 \leq |\gamma(Q_1, Q_2, \tau)| \leq 1. \tag{11}$$

The extreme value zero represents complete absence of correlation; the other extreme case, unity, represents complete correlation of the vibrations at Q_1 and Q_2. In the first case the

---

[1] This terminology is not quite uniform. What we call second-order and fourth-order coherence functions are sometimes called first-order and second-order coherence functions, respectively.

vibrations are said to be *mutually completely incoherent*; in the second case they are said to be *mutually completely coherent*.

A deeper insight into the significance of the complex degree of coherence may be obtained by expressing the interference law (8) in a somewhat different form, using the envelope representation of narrow-band signals which we discussed in Section 2.3. Let $\bar{\omega}$ be the mean frequency of the light and let us write

$$\gamma(Q_1, Q_2, \tau) = |\gamma(Q_2, Q_2, \tau)|e^{i[\alpha(Q_1, Q_2, \tau)-\bar{\omega}\tau]}, \tag{12}$$

where

$$\alpha(Q_1, Q_2, \tau) = \bar{\omega}\tau + \arg \gamma(Q_1, Q_2, \tau). \tag{13}$$

As we learned in Section 2.3, $\alpha(Q_1, Q_2, \tau)$ changes slowly over any time interval τ which is short relative to the reciprocal bandwidth $1/\Delta\omega$ of the light. If we express γ in the interference law (8) in the form (12), we obtain for the intensity at P the expression

$$I(P) = I^{(1)}(P) + I^{(2)}(P) + 2\sqrt{I^{(1)}(P)}\sqrt{I^{(2)}(P)} \; |\gamma(Q_1, Q_2, \tau)|\cos[\alpha(Q_1, Q_2, \tau) - \delta], \tag{14}$$

where

$$\delta = \bar{\omega}\tau = \bar{\omega}(t_2 - t_1) = \frac{2\pi}{\lambda}(R_2 - R_1), \tag{15}$$

$\lambda = 2\pi c/\bar{\omega}$ being the mean wavelength.

The formula (14) is an alternative form of the interference law (8) for stationary optical fields. Let us examine its implications.

Suppose that P_0 is some fixed point in the detection plane $\mathcal{B}$. The average intensities $I^{(1)}(P)$ and $I^{(2)}(P)$ will change slowly with P in the neighborhood of P_0 and so will $|\gamma(Q_1, Q_2, \tau)|$. Now, we have already noted that the phase factor $\alpha(Q_1, Q_2, \tau)$ will vary slowly with the time delay $\tau = (R_2 - R_1)/c$ over τ-intervals which are short compared with the coherence time of the light. Consequently, considered as a function of the phase delay δ, the last term in Eq. (14) will be a slowly modulated cosine term whose amplitude remains effectively constant over regions of the detection plane $\mathcal{B}$ for which the change in $R_2 - R_1$ is small in comparison with the coherence length of the light. Under these circumstances interference fringes will be formed in the observation plane $\mathcal{B}$, which will be essentially sinusoidal, with the amplitude and the phase of the sinusoidal patterns changing very slowly with position. This situation is illustrated in Fig. 3.2 for the commonly occurring situation when $I^{(2)}(P) = I^{(1)}(P)$. In this case the interference law (14) reduces to

$$I(P) = 2I^{(1)}(P)\{1 + |\gamma(Q_1, Q_2, \tau)|\cos[\alpha(Q_1, Q_2, \tau) - \delta]\}. \tag{16}$$

This formula implies that when $|\gamma| = 1$ (and only in this case) there are points in the fringe pattern where the average intensity is zero [namely points for which $\alpha(Q_1, Q_2, \tau) - \delta = (n + \frac{1}{2})\pi$ $(n = 0, \pm 1, \pm 2, \ldots)$], i.e. where there is complete cancellation of the light by

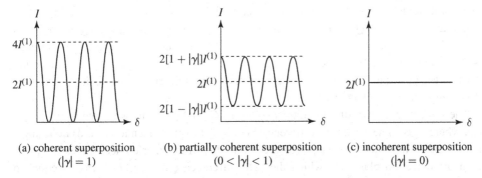

(a) coherent superposition
$(|\gamma| = 1)$

(b) partially coherent superposition
$(0 < |\gamma| < 1)$

(c) incoherent superposition
$(|\gamma| = 0)$

Fig. 3.2 The distribution of the average intensity in the neighborhood of an arbitrary point in the interference pattern formed by two quasi-monochromatic light beams of equal intensity $I^{(1)}$, with the correlation between them characterized by a degree of coherence γ.

destructive interference. This is the case of *complete coherence*. In the other extreme case, $\gamma = 0$, no interference fringes are formed at all, this being the case of *complete incoherence*. In all other cases, $(0 < |\gamma| < 1)$ fringes are formed but their contrast is lower than in the completely coherent case. One then says that the light at the pinholes is *partially coherent*.

It is evident from Eq. (16) that the maxima and minima of the average intensity in the immediate neighborhood of any point P in the plane $\mathcal{B}$ of observation are given by the formulas

$$I_{\max}(P) = 2I^{(1)}(P)[1 + |\gamma(Q_1, Q_2, \tau)|] \tag{17a}$$

and

$$I_{\min}(P) = 2I^{(1)}(P)[1 - |\gamma(Q_1, Q_2, \tau)|]. \tag{17b}$$

A useful measure of the contrast (i.e. of the sharpness) of the interference fringes is their so-called *visibility*, $\mathcal{V}$, defined as

$$\mathcal{V}(P) \equiv \frac{I_{\max}(P) - I_{\min}(P)}{I_{\max}(P) + I_{\min}(P)}. \tag{18}$$

On substituting from Eqs. (17) into Eq. (18) we see at once that

$$\mathcal{V}(P) = |\gamma(Q_1, Q_2, \tau)|. \tag{19}$$

This formula relates a quantity that can be readily measured (the visibility $\mathcal{V}$) to the more abstract concept of the modulus of the complex degree of coherence. The argument (the phase) of γ can also be determined experimentally, by measurement of the location of the intensity maxima in the fringe pattern (B&W, Section 10.4.1).

The complex degree of coherence $\gamma(Q_1, Q_2, \tau)$ provides a quantitative characterization of both temporal and spatial coherence, which were introduced in Sections 1.2 and 1.3 as two distinct concepts. The *temporal coherence* of light at one point Q is now characterized quantitatively by $\gamma(Q, Q, \tau)$. For example, in the Michelson interferometer, Q is a point on the dividing mirror M_0, shown in Fig. 1.2, and τ represents the time delay introduced between the two interfering beams by displacing one of the mirrors, M_1 or M_2, from the "symmetric position" by a distance $c\tau/2$. The visibility of the fringes in the detection plane $\mathcal{B}$ is then equal to $|\gamma(Q, Q, \tau)|$.

The *spatial coherence* of light at the two points Q_1 and Q_2 is characterized by $\gamma(Q_1, Q_2, \tau_0)$, where τ_0 is some fixed time difference $\tau_0 = t_2 - t_1$ (often taken to be zero) in Young's interference experiment. As we just saw, the visibility of the interference fringes at a point P in the detection plane, for which the path difference $\overline{Q_2P} - \overline{Q_1P} = c\tau_0$, is equal to $|\gamma(Q_1, Q_2, \tau_0)|$.

We will learn later (in Section 3.5) that only in special situations can temporal and spatial coherence of an optical field be treated independently of each other, because the mutual coherence function propagates according to precise differential equations, which relate its spatial and its temporal behavior.

Returning to spatial coherence, we have already mentioned that the time delay τ is often taken to be zero. This is generally the case when there is some kind of symmetry in the experimental set-up or when one is dealing with optical images near the axis of a centered optical system. In such cases the coherence effects are usually adequately described by the simpler correlation functions

$$J(Q_1, Q_2) = \Gamma(Q_1, Q_2, 0) = \langle V^*(Q_1, t)V(Q_2, t)\rangle \tag{20}$$

and

$$j(Q_1, Q_2) = \gamma(Q_1, Q_2, 0) = \frac{\Gamma(Q_1, Q_2, 0)}{\sqrt{\Gamma(Q_1, Q_1, 0)}\ \sqrt{\Gamma(Q_2, Q_2, 0)}} \tag{21a}$$

$$= \frac{J(Q_1, Q_2)}{\sqrt{J(Q_1, Q_1)}\ \sqrt{J(Q_2, Q_2)}} \tag{21b}$$

$$= \frac{J(Q_1, Q_2)}{\sqrt{I(Q_1)}\ \sqrt{I(Q_2)}}, \tag{21c}$$

where $I(Q_j) = \langle V^*(Q_j, t)V(Q_j, t)\rangle = J(Q_j, Q_j)$ ($j = 1, 2$) denotes the average intensity at the point Q_j. These functions may be called *equal-time coherence functions*. Often $J(Q_1, Q_2)$ is referred to as the *mutual intensity* and $j(Q_1, Q_2)$ as the *equal-time complex degree of coherence*.

It follows from the envelope properties of narrow-band light (discussed in Section 2.3) that

$$\Gamma(Q_1, Q_2, \tau) \approx J(Q_1, Q_2)e^{-i\bar{\omega}\tau}, \tag{22}$$

$$\gamma(Q_1, Q_2, \tau) \approx j(Q_1, Q_2)e^{-i\bar{\omega}\tau}, \tag{23}$$

provided that $|\tau|$ is small relative to the coherence time, i.e. provided that

$$|\tau| \ll \frac{2\pi}{\Delta\omega}. \tag{24}$$

3.2 Generation of spatial coherence from an incoherent source. The van Cittert–Zernike theorem

We have learned from our elementary treatments of the concepts of coherence area and of the coherence volume (Sections 1.3 and 1.4) that even a spatially incoherent source will generate a field which may be spatially coherent over large regions of space. Evidently in such situations spatial coherence is generated in the process of propagation. We will now give a simple intuitive argument indicating why this happens and we will then discuss this phenomenon quantitatively.

Consider light emitted by two small steady-state sources S_1 and S_2. We assume that the light is quasi-monochromatic of mean frequency $\bar{\omega}$ and that the sources are statistically independent, so that there is no correlation between the beams of light which they generate. We will compare the light vibrations at points some distance away from them.

Let $V_1(P_1, t)$ and $V_1(P_2, t)$ represent the field at points P_1 and P_2, respectively, due to the source S_1 and let $V_2(P_1, t)$ and $V_2(P_2, t)$ represent the field at these points due to the source S_2 (Fig. 3.3). If the difference between the distances $R_{11} = \overline{S_1P_1}$ and $R_{12} = \overline{S_1P_2}$ is small compared with the coherence length ($\sim 2\pi c/\Delta\omega$) of the light, one obviously has

$$V_1(P_2, t) \approx V_1(P_1, t). \tag{1a}$$

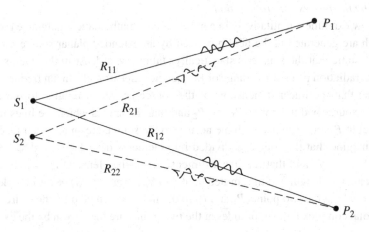

Fig. 3.3 Schematic illustration of the generation of spatial coherence from two small, uncorrelated sources.

Similarly, if the difference between the distances $R_{21} = \overline{S_2P_1}$ and $R_{22} = \overline{S_2P_2}$ is small relative to the coherence length of the light one has

$$V_2(P_2, t) \approx V_2(P_1, t). \tag{1b}$$

The total field at P_1 arises from the superposition of the field generated by the two sources (see Fig. 3.3) and hence is given by the expression

$$V(P_1, t) = V_1(P_1, t) + V_2(P_1, t). \tag{2a}$$

Similarly, the total field at P_2 is given by the expression

$$V(P_2, t) = V_1(P_2, t) + V_2(P_2, t). \tag{2b}$$

Now, since $V_1(P_1, t)$ and $V_2(P_1, t)$ are generated by sources S_1 and S_2 which are statistically independent, these two disturbances will not be correlated. For the same reason $V_1(P_2, t)$ and $V_2(P_2, t)$ will also not be correlated. However, the sums $V_1(P_1, t) + V_2(P_1, t)$ and $V_1(P_2, t) + V_2(P_2, t)$ will be correlated because of the relations (1) and, consequently,

$$V(P_2, t) \approx V(P_1, t). \tag{3}$$

This conclusion is also evident from the sketch in Fig. 3.3, where the (essentially identical) wave trains $V_1(P_1, t)$ and $V_1(P_2, t)$ arriving at the points P_1 and P_2, respectively, from S_1 are drawn as solid lines and the (essentially identical) wave trains $V_2(P_1, t)$ and $V_2(P_2, t)$ arriving at P_1 and P_2 from the other source S_2 are drawn as dashed lines. Clearly, although the solidly drawn wave trains and the wave trains drawn as dashed lines may have completely different forms, the *sum* of the two wave trains arriving at P_1 and the *sum* of the two wave trains arriving at P_2 will be similar to each other. Thus the fields at P_1 and P_2, given by Eqs. (2a) and (2b), will evidently be strongly correlated (i.e. statistically similar). Hence we see that, *even though the two small sources S_1 and S_2 are statistically independent, they will give rise to correlations in the field which they produce and the correlations are evidently generated by the process of propagation.*

Let us now examine quantitatively in a more precise mathematical language the correlations which are generated in the field produced by an extended planar source σ of natural light. We assume that the source is statistically stationary, at least in the wide sense, and that it emits radiation in a narrow range of frequencies $\Delta\omega$ around a mean frequency $\bar{\omega}$. We also assume that the linear dimensions of the source are small relative to the distances between the source and the points P_1 and P_2 and that the angles which the lines from each source point to P_1 and P_2 make with the normal to the source are small (see Fig. 3.4).

Let us imagine that the source is divided into elements $d\sigma_1$, $d\sigma_2$, ..., $d\sigma_m$ centered at points S_1, S_2, ..., S_m and that the linear dimensions of the elements are small compared with the mean wavelength $\bar{\lambda}$ of the emitted light. Let $V_{m1}(t)$ and $V_{m2}(t)$ be the complex amplitudes of the field at the two points P_1 and P_2 in the field, contributed by the source element $d\sigma_m$. The total complex field amplitudes at the two points are then given by the expressions

$$V_1(t) = \sum_m V_{m1}(t), \qquad V_2(t) = \sum_m V_{m2}(t). \tag{4}$$

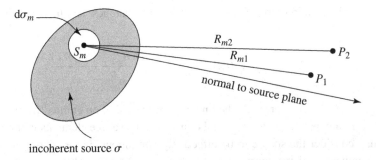

Fig. 3.4 Notation relating to the derivation of the van Cittert–Zernike theorem.

It follows that the mutual intensity $J(Q_1, Q_2)$, defined by Eq. (20) of Section 3.1, is, in the present case, given by the expression

$$
\begin{aligned}
J(P_1, P_2) &\equiv \langle V_1^*(t)V_2(t)\rangle \\
&= \sum_m \langle V_{m1}^*(t)V_{m2}(t)\rangle + \sum\sum_{m\neq n} \langle V_{m1}^*(t)V_{n2}(t)\rangle.
\end{aligned}
\tag{5}
$$

Since we assumed that the source generates natural light, the contributions from the different source elements may be assumed to be uncorrelated (mutually incoherent) and also of zero mean and hence

$$
\sum\sum_{m\neq n} \langle V_{m1}^*(t)V_{n2}(t)\rangle = \sum\sum_{m\neq n} \langle V_{m1}^*(t)\rangle\langle V_{n2}(t)\rangle = 0.
\tag{6}
$$

If R_{m1} and R_{m2} are the distance of the field points P_1 and P_2 from the source element $d\sigma_m$, then

$$
\left.
\begin{aligned}
V_{m1}(t) &= A_m(t - R_{m1}/c)\frac{\exp[-i\bar\omega(t - R_{m1}/c)]}{R_{m1}}, \\
V_{m2}(t) &= A_m(t - R_{m2}/c)\frac{\exp[-i\bar\omega(t - R_{m2}/c)]}{R_{m2}},
\end{aligned}
\right\}
\tag{7}
$$

where the modulus $|A_m|$ of A_m represents the strength and $\arg A_m$ represents the phase of the radiation emitted from the element $d\sigma_m$ of the source, c denoting the speed of light in vacuum.

On substituting from Eq. (7) into Eq. (5) and using Eq. (6), it follows that

$$
\begin{aligned}
J(P_1, P_2) &= \sum_m \langle A_m^*(t - R_{m1}/c)A_m(t - R_{m2}/c)\rangle \frac{\exp[i\bar\omega(R_{m2} - R_{m1})/c]}{R_{m1}R_{m2}} \\
&= \sum_m \langle A_m^*(t)A_m[t - (R_{m2} - R_{m1})/c]\rangle \frac{\exp[i\bar\omega(R_{m2} - R_{m1}/c]}{R_{m1}R_{m2}},
\end{aligned}
\tag{8}
$$

where we have used the fact that the source is statistically stationary. If the path differences $|R_{m2} - R_{m1}|$ are small relative to the coherence length of the light, $\Delta t \sim 2\pi/\Delta\omega$, we may also neglect the retardation terms $(R_{m2} - R_{m1})/c$ in the averages, and the formula (8) then becomes

$$J(P_1, P_2) = \sum_m \langle A_m^*(t) A_m(t) \rangle \frac{\exp[-i\bar{\omega}(R_{m2} - R_{m1})/c]}{R_{m1} R_{m2}}. \tag{9}$$

The average $\langle A_m^*(t) A_m(t) \rangle$ represents the intensity of the radiation emitted from the source element $d\sigma_m$. In any practical case the total number of the source elements will be so large that one may consider the source to be effectively continuous. If we denote the intensity emitted per unit area of the source by $I(S)$, then $\langle A_m^*(t) A_m(t) \rangle \approx I(S_m) d\sigma_m$, and we may replace the summation in Eq. (9) by integration. We then obtain the following expression for the mutual intensity at the pair of points P_1 and P_2:

$$J(P_1, P_2) = \int_\sigma I(S) \frac{e^{i\bar{k}(R_2 - R_1)}}{R_1 R_2} dS, \tag{10}$$

where R_1 and R_2 are the distances between a typical source point S and the field points P_1 and P_2 respectively, $\bar{k} = \bar{\omega}/c = 2\pi/\bar{\lambda}$ being the mean wave number of the light.

If we recall the definition Eq. (21c) of Section 3.1 of the equal-time degree of coherence $j(P_1, P_2)$ and use Eq. (10), we obtain at once the following formula for the equal-time degree of coherence of the light at the two field points:

$$j(P_1, P_2) = \frac{1}{\sqrt{I(P_1)}\sqrt{I(P_2)}} \int_\sigma I(S) \frac{e^{i\bar{k}(R_2 - R_1)}}{R_1 R_2} dS, \tag{11}$$

where

$$I(P_j) = J(P_j, P_j) = \int_\sigma \frac{I(S)}{R_j^2} dS \quad (j = 1, 2) \tag{12}$$

being the (averaged) intensities at P_1 and P_2.

The formula (11) is the mathematical formulation of a central theorem of optical coherence theory, known as the *van Cittert–Zernike theorem*. It expresses the equal-time degree of coherence $j(P_1, P_2)$ at two points P_1 and P_2 in a field generated by a planar, statistically stationary, spatially incoherent quasi-monochromatic source σ in terms of the averaged intensity distribution $I(S)$ across the source and the average intensities $I(P_1)$ and $I(P_2)$ at the two field points.

The integral that appears on the right of Eq. (11) is of the same form as is encountered in quite a different connection, namely in the theory of diffraction at an aperture in an opaque screen. To see this analogy we recall that if a monochromatic spherical wave

$$V(S, t) = U(S) e^{-i\omega t}, \tag{13a}$$

with

$$U(S) = a(S) \frac{e^{-ikR_1}}{R_1}, \tag{13b}$$

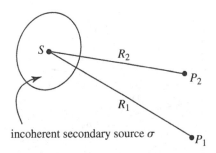

Coherence

van Cittert–Zernike Theorem

incoherent secondary source σ

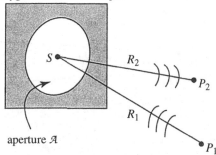

Diffraction at an Aperture

Huygens–Fresnel Principle

aperture $\mathcal{A}$

Equal-time degree of coherence of the field at points P_1 and P_2:

$$j(P_1, P_2) = \frac{1}{N} \int_\sigma I(S) \frac{e^{i\bar{k}(R_2 - R_1)}}{R_1 R_2}\, dS$$

Normalized complex amplitude $U(P_2)$ of the field at P_2 due to the diffraction of a monochromatic spherical wave converging to the point P_1:

$$U(P_2) = \frac{1}{N'} \int_{\mathcal{A}} a(S) \underbrace{\frac{e^{-ikR_1}}{R_1}}_{\substack{\text{incident} \\ \text{converging} \\ \text{wave}}} \underbrace{\frac{e^{ikR_2}}{R_2}}_{\substack{\text{diverging} \\ \text{spherical} \\ \text{secondary} \\ \text{wave}}}\, dS$$

Fig. 3.5 The analogy between the van Cittert–Zernike theorem and the Huygens–Fresnel principle.

converging to a point P_1 (see Fig. 3.5) is incident on an aperture $\mathcal{A}$ in an opaque screen, then the diffracted field at a point P_2 is, according to the Huygens–Fresnel principle [B&W, Sections 8.2 and 8.3, especially Eqs. (1) of Section 8.2 and Eq. (17) of Section 8.3], given by the expression (with the time-dependent factor $e^{-i\omega t}$ omitted)

$$U(P_2) = \frac{1}{N'} \int_{\mathcal{A}} a(S) \frac{e^{-ikR_1}}{R_1} \frac{e^{ikR_2}}{R_2}\, dS, \tag{14}$$

where N' is a constant, small angles of incidence and diffraction being assumed.

Comparison of the van Cittert–Zernike theorem expressed by the formula (11) with the Huygens–Fresnel principle (14) brings into evidence the following rather remarkable analogy: the van Cittert–Zernike theorem implies that, under the conditions stated, *the equal-time degree of coherence $j(P_1, P_2)$ is given by the normalized complex amplitude at a point P_2 in a certain diffraction pattern; namely that formed by a monochromatic spherical wave of frequency $\bar{\omega}$, converging towards a point P_1 and diffracted at an aperture $\mathcal{A}$ in an opaque screen of the same size, shape and location as the incoherent source σ, with the amplitude distribution across the aperture being proportional to the intensity distribution across the source.* (See Fig. 3.5.)

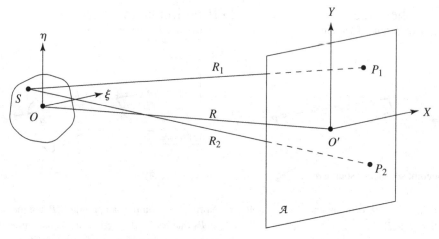

Fig. 3.6 Notation relating to the derivation of the far-zone form of the van Cittert–Zernike theorem.

We obtained this analogy from a formal comparison of the van Cittert–Zernike theorem of coherence theory with the Huygens–Fresnel principle of diffraction theory. A deeper reason for this analogy will become evident later (Section 3.5).

Frequently the field points P_1 and P_2 are located in the far zone of the source, often in a plane $\mathcal{A}$ parallel to the source plane. The van Cittert–Zernike theorem then takes on a simpler form which we will now derive. For this purpose we choose a Cartesian coordinate system in the source plane and denote by (ξ, η) the coordinates of a source point S. We also take a coordinate system in the plane $\mathcal{A}$ with origin O' and with the X, Y-axes parallel to the ξ, η-axes (see Fig. 3.6). If (X_1, Y_1) and (X_2, Y_2) are the Cartesian coordinates of the field points P_1 and P_2, respectively, in the plane $\mathcal{A}$, then the distances $R_1 = \overline{SP_1}$ and $R_2 = \overline{SP_2}$ are evidently given by the expressions

$$R_1^2 = (X_1 - \xi)^2 + (Y_1 - \eta)^2 + R^2,$$
$$R_2^2 = (X_2 - \xi)^2 + (Y_2 - \eta)^2 + R^2, \tag{15}$$

so that

$$R_1 \approx R + \frac{(X_1 - \xi)^2 + (Y_1 - \eta)^2}{2R},$$
$$R_2 \approx R + \frac{(X_2 - \xi)^2 + (Y_2 - \eta)^2}{2R}. \tag{16}$$

On the right-hand side of Eqs. (16) we retained only the leading terms of the power-series expansions, which is justified if we assume that the points P_1 and P_2 are at distances from the axis OO' which are small relative to R. It follows from Eqs. (16) that

$$R_2 - R_1 \approx \frac{(X_2^2 + Y_2^2) - (X_1^2 + Y_1^2)}{2R} - \frac{(X_2 - X_1)\xi + (Y_2 - Y_1)\eta}{R}. \qquad (17)$$

In the denominator of the integrands in Eqs. (11) and (12) R_1 and R_2 may be replaced, to a good approximation, by R. It is also convenient to set

$$\frac{X_2 - X_1}{R} = p, \qquad \frac{Y_2 - Y_1}{R} = q \qquad (18)$$

and

$$\psi = \frac{\bar{k}[(X_2^2 + Y_2^2) - (X_1^2 + Y_1^2)]}{2R}. \qquad (19)$$

It follows that when the points P_1 and P_2 are in the far zone, Eq. (11) for the equal-time degree of coherence reduces to

$$j(P_1, P_2) = \frac{e^{-i\psi} \iint_\sigma I(\xi, \eta) e^{-ik(p\xi + q\eta)} \, d\xi \, d\eta}{\iint_\sigma I(\xi, \eta) d\xi \, d\eta}. \qquad (20)$$

This formula shows that, *when the linear dimensions of the source and the distance between P_1 and P_2 are small relative to the distance of these points from the source, the equal-time degree of coherence $j(P_1, P_2)$, apart from a phase factor, is equal to the normalized Fourier transform of the intensity distribution across the source.* We may refer to Eq. (20) as the *far-zone form of the van Cittert–Zernike theorem.* The phase ψ, defined by Eq. (19), has a simple interpretation. It represents, to a good approximation, the phase difference $\bar{k}[P_2'P_2 - P_1'P_1]$ between the field points P_1 and P_2, in the plane $\mathcal{A}$, and points P_1' and P_2' at the same height from the OO' axis as P_1 and P_2, located on the sphere centered at the origin O in the source plane and passing through the origin O' in the plane $\mathcal{A}$ (see Fig. 3.7).

In its mathematical structure the expression on the right of Eq. (20) resembles the well-known expression of elementary diffraction theory for Fraunhofer diffraction at an aperture (B&W, Section 8.3.3). This was to be expected in view of the analogy between the van Cittert–Zernike theorem and the Huygens–Fresnel principle, which we noted earlier.

As an example let us determine the equal-time degree of coherence of the far field produced by an incoherent, quasi-monochromatic circular source of radius a and of uniform intensity i_0. The formula (20) gives, for this case,

$$j(P_1, P_2) = e^{i\psi} \frac{\tilde{I}(\bar{k}p, \bar{k}q)}{\tilde{I}(0, 0)}, \qquad (21)$$

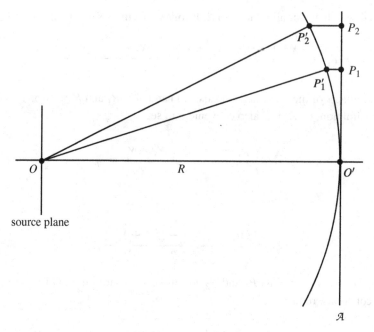

Fig. 3.7 Illustrating the significance of the phase ψ, defined by Eq. (19) of Section 3.2. It represents, to a good approximation, the phase difference $\bar{k}[\overline{P_2'P_2} - \overline{P_1'P_1}]$.

where

$$\tilde{I}(f, g) = i_0 \iint_{\xi^2 + \eta^2 \le a^2} e^{-i(f\xi + g\eta)} d\xi \, d\eta. \tag{22}$$

The integral on the right of Eq. (22) can readily be evaluated and one finds that (B&W, Section 8.5.2)

$$\tilde{I}(\bar{k}p, \bar{k}q) = \pi a^2 \left[\frac{2J_1\left(\bar{k}a\sqrt{p^2 + q^2}\right)}{\bar{k}a\sqrt{p^2 + q^2}} \right], \tag{23}$$

where J_1 is the Bessel function of the first kind and first order. On substituting from Eq. (23) into Eq. (21) we obtain for the equal-time degree of coherence of the far field the expression

$$j(P_1, P_2) = \frac{2J_1(v)}{v} e^{i\psi} \tag{24}$$

where

$$v = \bar{k}a\sqrt{p^2 + q^2}. \tag{25a}$$

Recalling the expressions for p and q given by Eqs. (18), we can express v in the form

$$v = \bar{k}\left(\frac{a}{R}\right)d, \tag{25b}$$

where

$$d = \sqrt{(X_1 - X_2)^2 + (Y_1 - Y_2)^2} \tag{26}$$

is the distance between the points P_1 and P_2 in the observation plane $\mathcal{A}$.

Apart from a trivial proportionality constant the expression on the right-hand-side of Eq. (24) will be recognized as the Airy formula for the field distribution in the Fraunhofer diffraction pattern of a uniformly coherently illuminated circular aperture [B&W, Section 8.5, Eq. (13)]. Its modulus is plotted in Fig. 3.8. It decreases steadily from the value unity when $v = 0$ to the value zero when $v = 3.83$ (indicated by the point B in the figure). Hence, with increasing separation of the two points P_1 and P_2 the equal-time degree of coherence decreases from the value unity (complete coherence) to zero (complete incoherence) when the right-hand side of Eq. (25b) is equal to 3.83, i.e. when

$$d = \frac{3.83}{\bar{k}}\left(\frac{R}{a}\right) = \frac{0.61R\bar{\lambda}}{a}. \tag{27}$$

A further increase in the separation of the two points reintroduces a small amount of coherence but the absolute value of the degree of coherence is then smaller than 0.14 and there is further incoherence, indicated by point C in Fig. 3.8 when $v = 7.02$.

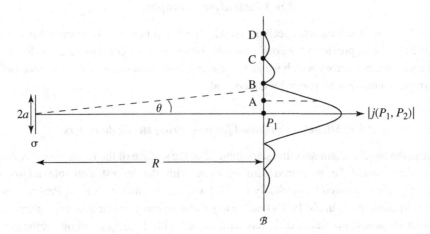

Fig. 3.8 The absolute value of the equal-time degree of coherence $j(P_1, P_2) = e^{i\psi}[2J_1(\bar{k}ad/R)/(\bar{k}ad/R)]$, [Eqs. (24) and (25b)], at points P_1 and P_2 in the far zone, generated by a uniform, incoherent, quasi-monochromatic circular source of radius a.

The function $2J_1(v)/v$ decreases steadily from the value unity when $v = 0$ to the value 0.88 when $v = 1$ (indicated by point A in the figure), i.e. for separation

$$d = \frac{0.16R\bar{\lambda}}{a}. \tag{28}$$

In practice, a drop from the value unity (complete coherence) which does not exceed 12% is often not regarded as being very significant. Hence, roughly speaking, *in the far zone and close to the direction normal to the source plane and parallel to it, the light produced by a spatially incoherent, quasi-monochromatic uniform, circular source of radius a is approximately coherent over a circular area* ΔA, *whose diameter is* $0.16\bar{\lambda}/\alpha$, *where* $\alpha = a/R$ *is the angular radius subtended by the source when viewed from* ΔA. We note that $\Delta A = \pi[0.16\bar{\lambda}/(2\alpha)]^2$, i.e.

$$\Delta A = 0.063 \frac{R^2}{S}\bar{\lambda}^2, \tag{29}$$

where $S = \pi a^2$ is the area of the source. This expression is in agreement with the order-of-magnitude relation (2) of Section 1.3 for the coherence area.

The far-zone behavior of the equal-time degree of coherence of light from an incoherent uniform, circular source which we just discussed was verified experimentally many years ago.[1] The results, together with the theoretical predictions, are shown in Fig. 3.9, for various separations of the pinholes. The experiment was repeated more recently, using a high-precision digital technique.[2] Excellent agreement with the theoretical predictions was obtained.

3.3 Illustrative examples

Two old classic interferometric techniques which we will now briefly discuss, both due to Albert Michelson, provide very good examples of some of the concepts and results of elementary coherence theory which we just discussed, even though they were invented before the notion of coherence came to be understood.

3.3.1 Michelson's method for measuring stellar diameters

Because the angular diameters that stars subtend at the surface of the Earth are exceedingly small, they cannot be measured directly even with the largest available telescopes. A. A. Michelson showed theoretically in 1890 and he, together with F. G. Pease, demonstrated experimentally in the 1920s that the angular diameter of a star and, in principle, also the intensity across the stellar disk may be obtained with the help of an interferometer as

[1] B. J. Thompson and E. Wolf, *J. Opt. Soc. Amer.* **47** (1957), 895–902.
[2] G. Ambrosini, G. Schirripa Spagnola, D. Paoletti and S. Vicalvi, *Pure Appl. Opt.* **7** (1989), 933–939.

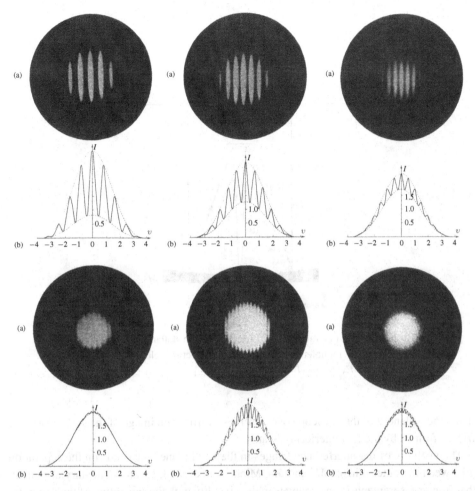

Fig. 3.9 Young's interference pattern formed by partially coherent quasi-monochromatic light with three different degrees of coherence, from a uniform incoherent circular source, for various separations of the pinholes. Upper figures: observed patterns. Lower figures: results of experiments. For details of the experimental arrangement see B&W Section 10.4.3 or the original paper by B. J. Thompson and E. Wolf, *J. Opt. Soc. Amer.* **47** (1957), 895–902, from which these figures are reproduced.

shown schematically in Fig. 3.10. The principle of the technique may be understood as follows. Light from the star is incident on the outer mirrors M_1 and M_2 of the interferometer, is then reflected at two inner mirrors M_3 and M_4 and is brought to the back focal plane $\mathcal{F}$ of a telescope to which the interferometer is attached. The purpose of the telescope is to provide stability for the interferometer. The inner mirrors M_3 and M_4 are fixed while the outer mirrors M_1 and M_2 can be separated symmetrically in the direction joining M_3 and M_4.

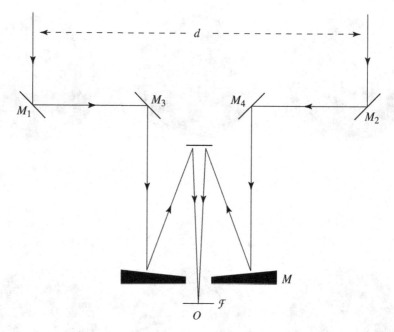

Fig. 3.10 A schematic diagram of the 20-foot Michelson stellar interferometer at Mount Wilson Observatory, mounted on the 100-inch reflecting telescope. [Adapted from F. G. Pease, *Ergeb. Ex. Naturwiss.* **10** (1931), 84–96.]

In the focal plane $\mathcal{F}$ of the telescope one observes a diffraction image of the star, crossed by fringes formed by the two interfering beams.

The visibility of the interference fringes in the focal plane $\mathcal{F}$ depends on the separation d between the outer mirrors M_1 and M_2. Michelson showed, by an elementary argument, that from measurements of the changes in the visibility with the separation of the two outer mirrors one may obtain information about the intensity distribution across the star, at least in cases when the star is assumed to be rotationally symmetric. Michelson showed, in particular, that, if the stellar disk is circular and uniform, the visibility curve, considered as a function of the separation of the two outer mirrors M_1 and M_2, will have zeros for certain separation distances d; and that for a spectral component of wavelength λ_0 the smallest of the values for which a zero occurs is

$$d_0 = \frac{0.61\lambda_0}{\alpha}, \tag{1}$$

where α is the angular radius of the star. Thus from measurement of d_0 the angular diameter of the star may be determined.

From the standpoint of coherence theory, the principle of the method can easily be understood. At the two outer mirrors M_1 and M_2, the incident light is, in general, partially coherent. According to the far zone of the van Cittert–Zernike theorem [Eq. (20) of Section 3.2; see also Fig. 3.6], expressed in a slightly different form, the equal-time degree of coherence is given by the formula (with $|\psi| \ll 1$, $k_0 = \omega_0/c = 2\pi/\lambda_0$)

$$j(M_1, M_2) = \frac{\iint_\sigma i(u, v)e^{ik_0[(x_2-x_1)u+(y_2-y_1)v]}\,du\,dv}{\iint_\sigma i(u, v)\,du\,dv}. \tag{2}$$

Here $i(u, v)$ represents the (averaged) intensity distribution across the stellar disk as a function of the angular variables u and v of the points (ξ, η) on that disk,

$$u = \frac{\xi}{R}, \qquad v = \frac{\eta}{R}, \tag{3}$$

and (x_1, y_1) and (x_2, y_2) are the coordinates of the outer mirrors M_1 and M_2.

It follows at once from the formula (2) and the fact that the visibility of the fringes is equal to the absolute value of the equal-time complex degree of coherence [Eqs. (23) and (19) of Section 3.1] that the visibility of the fringes in the focal plane of the telescope is proportional to the Fourier transform of the (averaged) intensity distribution across the star. In particular, if the stellar disk is circular, is of uniform intensity and subtends angular radius $\alpha = a/R$ at the telescope, the formula (2) gives [cf. Eq. (24) of Section 3.2]

$$|j(M_1, M_2)| = \left|\frac{2J_1(k_0\alpha d)}{k_0\alpha d}\right|, \tag{4}$$

where $d = \sqrt{(x_2 - x_1)^2 + (y_2 - y_1)^2}$ is the distance between the two outer mirrors M_1 and M_2 of the interferometer, J_1 being the Bessel function of the first kind and first order. The smallest value d_0 of d for which the expression (4) vanishes is given by $k_0\alpha d_0 = 3.83$, implying that $d_0 = 0.61\lambda_0/\alpha$, in agreement with Michelson's result expressed by Eq. (1).

The first determination of a stellar diameter using this technique was made in the 1920s and employed a 20-foot interferometer attached to the 100-inch telescope at the Mount Wilson Observatory (Fig. 3.11). By means of it the angular diameter of the red giant star Betelgeuse in the constellation of Orion was determined by F. G. Pease. In these measurements the first "incoherence" occurred when the outer mirrors M_1 and M_2 of the interferometer were separated by the distance of $d_0 = 307$ cm. According to the formula (1) this implies, taking $\overline{\lambda}_0 = 5.75 \times 10^{-5}$ cm, that the angular diameter of Betelgeuse is about 0.047 seconds of arc. Angular diameters of five other stars were determined by that interferometer in the 1920s, but the instrument has not been used since that time. However, it has been preserved at the Observatory. A photograph of it, taken in more recent times, is shown in Fig. 3.12.

Fig. 3.11 The Michelson stellar interferometer mounted on the 100-inch telescope. [Adapted from F. G. Pease, *Ergeb. Ex. Naturwiss* **10** (1931), 84–96.]

Fig. 3.12 The original Michelson stellar interferometer preserved at the Mount Wilson Observatory, around the year 2000. (Courtesy of Gale Gant and Don Nicholson.)

Fig. 3.13 The Westbrook aperture-synthesis telescope. (Reproduced from K. Rohlfs, *Tools of Radio Astronomy*, Springer, Berlin and New York, 1986, p. 113, Fig. 6.9.)

Since the time when the first Michelson stellar interferometer was built in the 1920s, several others have been constructed and used, one of them operating at infrared wavelengths. Today, however, this technique is used mainly in radio astronomy, where the principle has been applied with great success to map the radio sky. Because the wavelengths of radio waves are very much longer than those of light, the base line of the radio interferometer – usually called a radio telescope or an antenna-synthesis telescope – has to be many orders of magnitude longer. Instead of mirrors one uses large antennas (see Fig. 3.13) and the incoming radio waves are sampled at pairs of them. The antennas are arranged in various configurations, one of which is shown in Fig. 3.14.

3.3.2 Michelson's method for determining energy distribution in spectral lines

Suppose that a beam of quasi-monochromatic light is divided into two beams in a Michelson interferometer (Fig. 1.2) and that the beams are superposed after a path difference $c\tau$ has been introduced between them. In the region of superposition interference fringes whose visibility depends on the path difference are formed. Michelson showed in the 1890s that from measurements of the visibility $\mathcal{V}(\tau)$ of the interference fringes, as a function of τ one may obtain information about the energy distribution in the spectrum of the light.

Fig. 3.14 The VLA (Very Large Array) at Socorro, New Mexico. The array consists of movable 25-m telescopes and operates with wavelengths $\approx 3\,\text{cm}$. (Reproduced from K. Rohlfs, *Tools of Radio Astronomy*, Springer, Berlin and New York, 1986, p. 113, Fig. 6.9.)

From the standpoint of coherence theory the principle of this method may be understood as follows. Let us for simplicity assume that the two beams have the same intensity. Then, according to Eq. (19) of Section 3.1, the visibility of the interference fringes, (considered now as a function of τ rather than P), in the region of superposition of the two beams is given by the expression

$$\mathcal{V}(\tau) = |\gamma(\tau)|, \tag{1}$$

where $\gamma(\tau) = \gamma(Q_0, Q_0, \tau)$, Q_0 denoting a typical point on the dividing mirror M_0. According to the normalized form of the Wiener–Khintchine theorem [Eq. (9b) of Section 2.5],

$$\gamma(\tau) = \int_0^\infty s(\omega) e^{-i\omega\tau} \, d\omega, \tag{2}$$

where $s(\omega)$ is the normalized spectral density of the light at Q_0. The integral on the right contains only positive frequency components, because we use the analytic signal representation (which we discussed in Section 2.3) of the optical field.

It is convenient to set

$$\gamma(\tau) = \hat{\gamma}(\tau) e^{-i\omega_0\tau}, \tag{3}$$

where ω_0 is a mid-frequency of the light. It then follows from Eqs. (3) and (2) that

$$\hat{\gamma}(\tau) = \int_{-\infty}^{\infty} \hat{s}(\mu)e^{-i\mu\tau}\,d\mu, \tag{4}$$

where

$$\hat{s}(\mu) = \begin{cases} s(\omega_0 + \mu) & \text{when } \mu \geq -\omega_0, \\ 0 & \text{when } \mu < -\omega_0, \end{cases} \tag{5}$$

From Eqs. (1)–(5) it follows that

$$\mathcal{V}(\tau) = \left| \int_{-\infty}^{\infty} \hat{s}(\mu)e^{-i\mu\tau}\,d\mu \right|. \tag{6}$$

Suppose first that the spectrum is symmetric about the central frequency ω_0. Then the "shifted" spectrum $\tilde{s}(\mu)$ will be approximately an even function of μ and, consequently, the integral appearing in Eq. (6) will be real. It follows that in this case

$$\mathcal{V}(\tau) = \pm 2\int_{0}^{\infty} \hat{s}(\mu)\cos(\mu\tau)d\mu. \tag{7}$$

The ambiguity in the choice of the sign on the right-hand side of Eq. (6) arises from the fact that the expression (6) gives only the modulus of the integral. On taking the inverse of Eq. (7) we obtain for the shifted normalized spectrum the expression

$$\hat{s}(\mu) \equiv s(\omega_0 + \mu) = \pm\frac{1}{\pi}\int_{0}^{\infty} \mathcal{V}(\tau)\cos(\mu\tau)d\tau. \tag{8}$$

This formula shows that, when the spectrum is symmetric, the spectral energy distribution about the mid-frequency ω_0 may be calculated from measurements of the visibility curve, provided that the ambiguity in sign of the integral can be removed. This may usually be done by appealing to physical plausibility.

If the spectrum is not symmetric, the Fourier transform of the "shifted" spectral density $\hat{s}(\mu)$ is no longer everywhere real and in such cases Eq. (8) no longer applies. In order to determine the spectral density in such cases one needs to know, in addition to the visibility curve, also the phase of the Fourier transform $\hat{\gamma}(\tau)$ of $\hat{s}(\mu)$ or, alternatively, the phase of the complex degree of coherence. As mentioned earlier [in the paragraph which follows Eq. (19) of Section 3.1], the phase of the complex degree of coherence can be determined from measurements of the location of the intensity maxima in the fringe pattern formed by the interfering beams. In Fig. 3.15 a result from Michelson's original determination by this technique of the energy distribution in spectral lines of thalium is reproduced.

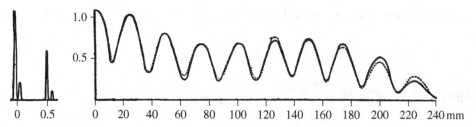

Fig. 3.15 Two lines, each having a satellite, in the spectrum of thallium (left) and the corresponding visibility curve (right). (Adapted from A. A. Michelson, *Light Waves and Their Uses*, The University of Chicago Press, Chicago, IL, 1902, Reproduced from first Phoenix Science Series, University of Chicago Press, 1961, Fig. 64, p. 79.)

In more recent times Michelson's method has been superseded by a related interferometric technique, called *Fourier spectroscopy*, which is also known as the *interferogram method*, and sometimes called the *FTIR technique*. It is used mainly in the infrared region of the spectrum. This method allows direct determination both of the real part and of the imaginary part of the degree of coherence $\gamma(\tau)$, from which the normalized spectral density may be unambiguously determined.

3.4 Propagation of the mutual intensity

The van Cittert–Zernike theorem that we discussed in Section 3.2 implies that the spatial coherence of light changes on propagation. Specifically the theorem indicates that even a spatially incoherent source generates a field that is partially coherent and is, in some region of space, highly coherent. We will now generalize this result to propagation from an open surface on which the equal-time degree of coherence is known.

Suppose that the complex amplitude at a typical point Q on a surface $\mathcal{A}$ intercepting a quasi-monochromatic light beam is $V(Q, t) \approx U(Q)\exp(-i\bar{\omega}t)$. Then, according to the Huygens–Fresnel principle (B&W, Section 8.2) the space-dependent part of the complex amplitude at a point P_1 in the domain in which the beam propagates is given by the expression (assuming small angles of incidence and diffraction)

$$U(P_1) = -\frac{i}{\lambda} \iint_{\mathcal{A}} U(Q_1)\frac{e^{i\bar{k}R_1}}{R_1}\,dQ_1, \tag{1}$$

where R denotes the distance from Q_1 to P_1 (see Fig. 3.16). Similarly,

$$U(P_2) = -\frac{i}{\lambda} \iint_{\mathcal{A}} U(Q_2)\frac{e^{i\bar{k}R_2}}{R_2}\,dQ_2, \tag{2}$$

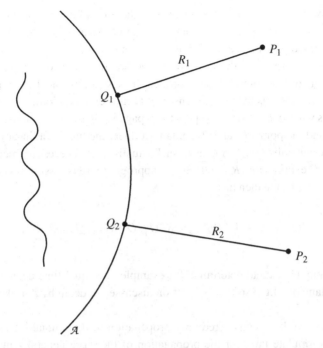

Fig. 3.16 Notation relating to derivation of Zernike's propagation law for the mutual intensity [Eq. (5) of Section 3.4].

$\bar{k} = \omega/c$ being the mean wave number associated with the mean frequency $\bar{\omega}$. It follows that

$$\langle U^*(P_1)U(P_2) \rangle = \frac{1}{\bar{\lambda}^2} \int_{\mathcal{A}} \int_{\mathcal{A}} \langle U^*(Q_1)U(Q_2) \rangle \frac{e^{i\bar{k}(R_2-R_1)}}{R_1 R_2} dQ_1 \, dQ_2. \tag{3}$$

The expectation values appearing in this formula will be recognized as the mutual intensity at the points Q_1 and Q_2 [cf. Eq. (20) of Section 3.1] and hence the formula (3) implies that

$$J(P_1, P_2) = \frac{1}{\bar{\lambda}^2} \int_{\mathcal{A}} \int_{\mathcal{A}} J(Q_1, Q_2) \frac{e^{i\bar{k}(R_2-R_1)}}{R_1 R_2} dQ_1 \, dQ_2. \tag{4}$$

This formula was first derived by F. Zernike in his basic paper on coherence cited earlier, published in 1938, and is often called *Zernike's propagation law* for the mutual intensity.

If we recall the definition of the equal-time degree of coherence $j(Q_1, Q_2)$ [Eq. (21c) of Section 3.1] we may express Zernike's propagation law (4) in the form

$$j(P_1, P_2) = \frac{1}{\sqrt{I(P_1)}\sqrt{I(P_2)}} \int_{\mathcal{A}} \int_{\mathcal{A}} j(Q_1, Q_2)\sqrt{I(Q_1)}\sqrt{I(Q_2)} \frac{e^{i\bar{k}(R_2-R_1)}}{R_1 R_2} dQ_1 \, dQ_2. \tag{5}$$

This formula expresses the equal-time degree of coherence of the light at any pair of points P_1 and P_2 in the region of space into which the light propagates in terms of the equal-time degree of coherence and the averaged intensity of the light across the surface $\mathcal{A}$. The formula (5) is a generalization of the van Cittert–Zernike theorem and reduces to it when the light distribution on the (now planar) surface $\mathcal{A}$ is spatially incoherent [$J(Q_1, Q_2) \sim \delta^{(2)}(Q_2 - Q_1)$, where $\delta^{(2)}$ is the two-dimensional Dirac delta function].

The formulas which we just derived apply to propagation in free space. If the space between the surface $\mathcal{A}$ and the points P_1 and P_2 contains a linear medium or a linear optical system one can readily generalize the formulas to such situations. It is necessary merely to replace the "propagator" $\exp(i\bar{k}R)/R$ ($R = \overline{QP}$) by an appropriate transmission function, say $K(P, Q)$. Instead of Eq. (4) one then has

$$J(P_1, P_2) = \int_{\mathcal{A}} J(Q_1, Q_2) K^*(P_1, Q_1) K(P_2, Q_2) dQ_1 \, dQ_2. \tag{6}$$

This formula may be used to determine, for example, the equal-time degree of coherence in the image plane of a light source, a situation discussed in detail by Zernike in his classic paper.

In this section we have considered only propagation of the mutual intensity $J(P_1, P_2)$. One can also formulate laws for the propagation of the more general mutual coherence function $\Gamma(P_1, P_2, \tau)$ which are somewhat more complicated. They may be derived from rigorous propagation laws for the mutual coherence function, to which we will now turn our attention.

3.5 Wave equations for the propagation of mutual coherence in free space

We saw in Section 3.2 that the mathematical formulas which express the van Cittert–Zernike theorem for the mutual intensity and the equal-time degree of coherence closely resemble a well-known formula of elementary diffraction theory, namely the Huygens–Fresnel principle. The propagation law for the mutual intensity which we just derived likewise resembles the Huygens–Fresnel principle. We will now show that there is a deeper reason for these analogies than might appear to be the case from our formal derivation.

Consider an ensemble $\{V(\mathbf{r}, t)\}$ representing a complex wavefield in free space.[1] Each member of the ensemble satisfies the wave equation

$$\nabla^2 V(\mathbf{r}, t) = \frac{1}{c^2} \frac{\partial^2 V(\mathbf{r}, t)}{\partial t^2}. \tag{1}$$

---

[1] In the preceding sections we frequently denoted points by capital letters, such as P, Q, and S. In the general theory it is, however, more convenient to represent them by position vectors such as $\mathbf{r}_1$, $\mathbf{r}_2$ etc. In the subsequent sections we will use either of the notations, whichever is more convenient.

Let us take the complex conjugate of this equation, replace $\mathbf{r}$ by $\mathbf{r}_1$ and t by t_1 and multiply the resulting equation by $V(\mathbf{r}_2, t_2)$. This gives

$$\nabla_1^2 V^*(\mathbf{r}_1, t_1) V(\mathbf{r}_2, t_2) = \frac{1}{c^2} \frac{\partial^2 V^*(\mathbf{r}_1, t_1)}{\partial t_1^2} V(\mathbf{r}_2, t_2), \tag{2}$$

where ∇_1^2 is the Laplacian operator acting with respect to the point $\mathbf{r}_1$. Next let us take the ensemble average of both sides of this equation and interchange the order of the various operations. We then obtain the equation

$$\nabla_1^2 \langle V^*(\mathbf{r}_1, t_1) V(\mathbf{r}_2, t_2) \rangle = \frac{1}{c^2} \frac{\partial^2}{\partial t_1^2} \langle V^*(\mathbf{r}_1, t_1) V(\mathbf{r}_2, t_2) \rangle. \tag{3}$$

If the field is statistically stationary, at least in the wide sense as we now assume, then

$$\langle V^*(\mathbf{r}_1, t_1) V(\mathbf{r}_2, t_2) \rangle = \langle V^*(\mathbf{r}_1, t) V(\mathbf{r}_2, t + t_2 - t_1) \rangle = \Gamma(\mathbf{r}_1, \mathbf{r}_2, \tau), \tag{4}$$

$\tau = t_2 - t_1$, where $\Gamma(\mathbf{r}_1, \mathbf{r}_2, \tau)$ is the mutual coherence function of the field [Eq. (6) of Section 3.1]. Evidently $\partial^2/\partial t_1^2 = \partial^2/\partial \tau^2$ and hence Eq. (3) implies that

$$\nabla_1^2 \Gamma(\mathbf{r}_1, \mathbf{r}_2, \tau) = \frac{1}{c^2} \frac{\partial^2 \Gamma(\mathbf{r}_1, \mathbf{r}_2, \tau)}{\partial \tau^2}. \tag{5a}$$

In a strictly similar manner we find that

$$\nabla_2^2 \Gamma(\mathbf{r}_1, \mathbf{r}_2, \tau) = \frac{1}{c^2} \frac{\partial^2 \Gamma(\mathbf{r}_1, \mathbf{r}_2, \tau)}{\partial \tau^2}, \tag{5b}$$

where ∇_2^2 is the Laplacian operator acting with respect to the point $\mathbf{r}_2$.

The two wave equations[1] (5a) and (5b) for the mutual coherence function hold rigorously for propagation in free space. From them one can obtain at once equations which are valid, to a good approximation, for the propagation of the mutual intensity in free space. They follow on substituting for Γ from Eq. (22) of Section 3.1 into the wave equations (5). One then obtains the two equations

$$\nabla_1^2 J(\mathbf{r}_1, \mathbf{r}_2) + \bar{k}^2 J(\mathbf{r}_1, \mathbf{r}_2) \approx 0, \tag{6a}$$

and

$$\nabla_2^2 J(\mathbf{r}_1, \mathbf{r}_2) + \bar{k}^2 J(\mathbf{r}_1, \mathbf{r}_2) \approx 0, \tag{6b}$$

where

$$\bar{k} = \frac{\bar{\omega}}{c}, \tag{7}$$

---

[1] Actually the two equations (5) are not independent, because of the relation $\Gamma(\mathbf{r}_1, \mathbf{r}_2, \tau) = \Gamma^*(\mathbf{r}_2, \mathbf{r}_1, -\tau)$.

provided that the inequality given by Eq. (24) of Section 3.1 holds, i.e. provided that

$$|\tau| \ll \frac{2\pi}{\Delta\omega}. \tag{8}$$

Equations (6) explain why the van Cittert–Zernike theorem resembles closely the Huygens–Fresnel principle of elementary diffraction theory. This principle is a consequence of the fact that a monochromatic optical field satisfies the Helmholtz equation. We have just shown that the mutual intensity of a stationary random field also obeys, approximately, the Helmholtz equation. This common property implies that in both cases the propagation is, to a good approximation, governed by the Huygens–Fresnel principle.

PROBLEMS

3.1 A plane, polychromatic light wave, whose spectrum consists of a line with Gaussian profile, is incident normally on a screen $\mathcal{A}$ containing two pinholes. At each instant of time the complex wave amplitudes at the two pinholes are the same. Derive an expression for the visibility of the interference fringes observed at a point P on a screen $\mathcal{B}$ parallel to $\mathcal{A}$, at distances $\mathbf{r}_1$ and $\mathbf{r}_2$ from the pinholes.

3.2 A double star consists of two components which subtend on Earth the same angular diameters 2α and have angular separation 2β. The stars may be regarded as having uniform circular cross-sections and radiating at the same mean wavelength. The ratio of the brightnesses of the two components is $1 : b$. The light received from the star is passed through a filter so as to make it quasi-monochromatic.
   (a) Derive an expression for the equal-time degree of coherence j_{12} of the star light in the observing plane of a Michelson stellar interferometer.
   (b) If $\beta \gg \alpha$, show how β may be determined from the visibility curve.

3.3 A Michelson interferometer is illuminated by a quasi-monochromatic light beam having a rectangular spectral distribution of width $\Delta\omega$, centered on frequency $\bar{\omega}$. The interference fringes first vanish when one of the mirrors is displaced by a distance d_0 from its symmetric position with respect to the other mirror. Determine the degree of self-coherence $\gamma(\mathbf{r}, \mathbf{r}, \tau)$ of the light at the beam splitter of the interferometer and using it, calculate the width $\Delta\omega$ of the spectral distribution.

3.4 A spectrum of light consists of N "lines" of the same profile but of different intensities, centered at frequencies $\omega_1, \omega_2, \ldots, \omega_N$. The light is analyzed by means of a Michelson interferometer.
   (a) Derive an expression for the visibility curve.
   (b) Discuss in detail the case when the spectrum consists of two lines ($N = 2$) of the same intensities and with identical profiles that are of Gaussian form. Show also how the separation of the two lines may be determined if it is assumed to be large relative to the effective width of each line.

3.5 The mutual intensity function for all pairs of points on a surface intercepting a stationary quasi-monochromatic light beam is of the form

$$J(\mathbf{r}_1, \mathbf{r}_2) = f(\mathbf{r}_1)g(\mathbf{r}_2),$$

where $f(\mathbf{r})$ and $g(\mathbf{r})$ are known functions of position on the surface. Show that
(a) $g(\mathbf{r}) = \alpha f^*(\mathbf{r})$, where α is a real constant and the asterisk denotes the complex conjugate; and
(b) the light in the space into which it propagates is completely spatially coherent within the framework of second-order coherence theory.

3.6 Consider a real random source-distribution $Q^{(r)}(\mathbf{r}, t)$, localized for all times within a finite volume D in free space, and let $V^{(r)}(\mathbf{r}, t)$ be the field which the source generates. $Q^{(r)}$ and $V^{(r)}$ are related by the inhomogeneous wave equation

$$\nabla^2 V^{(r)}(\mathbf{r}, t) - \frac{1}{c^2}\frac{\partial^2 V^{(r)}(\mathbf{r}, t)}{\partial t^2} = -4\pi Q^{(r)}(\mathbf{r}, t).$$

Show that if $Q^{(r)}(\mathbf{r}, t)$ and $V^{(r)}(\mathbf{r}, t)$ are stationary random processes, and $Q(\mathbf{r}, t)$ and $V(\mathbf{r}, t)$ are the associated analytic signals, then the cross-correlation functions

$$\Gamma_Q(\mathbf{r}_1, \mathbf{r}_2, \tau) = \langle Q^*(\mathbf{r}_1, t)Q(\mathbf{r}_2, t + \tau)\rangle$$

and

$$\Gamma_V(\mathbf{r}_1, \mathbf{r}_2, \tau) = \langle V^*(\mathbf{r}_1, t)V(\mathbf{r}_2, t + \tau)\rangle$$

are related by the equation

$$\left(\nabla_2^2 - \frac{1}{c^2}\frac{\partial^2}{\partial \tau^2}\right)\left(\nabla_1^2 - \frac{1}{c^2}\frac{\partial^2}{\partial \tau^2}\right)\Gamma_V(\mathbf{r}_1, \mathbf{r}_2, \tau) = (4\pi)^2\Gamma_Q(\mathbf{r}_1, \mathbf{r}_2, \tau).$$

3.7 The mutual coherence function of a certain stationary optical field in free space has the form

$$\Gamma(\mathbf{r}_1, \mathbf{r}_2, \tau) = F(\mathbf{r}_1, \mathbf{r}_2)G(\tau).$$

Show that the function $F(\mathbf{r}_1, \mathbf{r}_2)$ must satisfy two Helmholtz equations and determine the most general form of $G(\tau)$.

4

Second-order coherence phenomena in
the space–frequency domain

Up to relatively recent times, second-order coherence phenomena have usually been described in terms of the space–time correlation function, namely the mutual coherence function $\Gamma(\mathbf{r}_1, \mathbf{r}_2, \tau)$, or in terms of a space correlation function, i.e. the mutual intensity $J(\mathbf{r}_1, \mathbf{r}_2)$. Using them we defined, in Section 3.1, the complex degrees of coherence $\gamma(\mathbf{r}_1, \mathbf{r}_2, \tau)$ and $j(\mathbf{r}_1, \mathbf{r}_2)$. More recently an alternative description was developed, which has considerable advantages in the analysis of many problems involving statistical wavefields. It employs certain functions which were originally introduced rather formally in terms of the Fourier transforms of the mutual coherence function, but later they were found to be also correlation functions, associated with ensembles of realizations, that are functions of position and frequency, rather than of position and time. This step is not as trivial as it might appear at first sight because, as we noted in Section 2.5, the sample functions of a stationary random process do not have a Fourier frequency representation.

The newer space-frequency representation turned out to be very useful for providing solutions to many problems and it has led to the discovery of some new effects, some of which we will discuss in this chapter.

4.1 Coherent-mode representation and the cross-spectral density
as a correlation function

As we already noted, in the space–time formulation of coherence theory of stationary optical fields, the basic quantity is the mutual coherence function

$$\Gamma(\mathbf{r}_1, \mathbf{r}_2, \tau) = \langle V^*(\mathbf{r}_1, t) V(\mathbf{r}_2, t + \tau) \rangle. \tag{1}$$

In the space–frequency formulation, the basic quantity is the cross-spectral density function $W(\mathbf{r}_1, \mathbf{r}_2, \omega)$, which is its Fourier frequency transform, i.e.

$$W(\mathbf{r}_1, \mathbf{r}_2, \omega) = \frac{1}{2\pi} \int_{-\infty}^{\infty} \Gamma(\mathbf{r}_1, \mathbf{r}_2, \tau) e^{i\omega\tau} \, d\tau. \tag{2}$$

For later purposes we note that in free space the cross-spectral density obeys the two Helmholtz equations

$$\nabla_1^2 W(\mathbf{r}_1, \mathbf{r}_2, \omega) + k^2 W(\mathbf{r}_1, \mathbf{r}_2, \omega) = 0 \tag{3a}$$

and

$$\nabla_2^2 W(\mathbf{r}_1, \mathbf{r}_2, \omega) + k^2 W(\mathbf{r}_1, \mathbf{r}_2, \omega) = 0, \tag{3b}$$

where ∇_1^2 and ∇_2^2 are the Laplacian operators acting with respect to the points $\mathbf{r}_1$ and $\mathbf{r}_2$, respectively, $k = \omega/c$ is the free-space wave number, c being the speed of light in vacuum. These equations follow at once from the two wave equations (5a) and (5b) of Section 3.5 which are satisfied by the mutual coherence function in free space, on taking their Fourier transforms.

We have already encountered the cross-spectral density in a somewhat broader context, in connection with the generalized Wiener–Khintchine theorem [Eqs. (12a) and (12b) of Section 2.5] of the theory of stationary random processes. However, in that treatment the cross-spectral density appeared in a "singular formula" containing the Dirac delta function [Eq. (11) of Section 2.5]. In the space–frequency formulation of coherence theory, the cross-spectral density is introduced in an alternative way, within the framework of ordinary function theory, as a correlation function of a statistical ensemble of well-behaved realizations.

Let us consider an optical field in a closed domain D in free space. Then it can be shown (M&W, Sections 4.7.1 and 4.7.2) that, under very general conditions (Hermiticity, non-negative definiteness and square-integrability of W over D), the cross-spectral density of the field at any pair of points $\mathbf{r}_1$ and $\mathbf{r}_2$ in D may be expressed in a (generally infinite) series

$$W(\mathbf{r}_1, \mathbf{r}_2, \omega) = \sum_n \lambda_n(\omega) \phi_n^*(\mathbf{r}_1, \omega) \phi_n(\mathbf{r}_2, \omega). \tag{4}$$

The functions ϕ_n may be shown to be the eigenfunctions and λ_n the eigenvalues of the integral equation

$$\int_D W(\mathbf{r}_1, \mathbf{r}_2, \omega) \phi_n(\mathbf{r}_1, \omega) \mathrm{d}^3 r_1 = \lambda_n(\omega) \phi_n(\mathbf{r}_2, \omega). \tag{5}$$

The eigenfunction ϕ_n may be taken to form an orthonormal set over the domain D, i.e.

$$\int_D \phi_n^*(\mathbf{r}, \omega) \phi_m(\mathbf{r}, \omega) \mathrm{d}^3 r = \delta_{nm}, \tag{6}$$

δ_{nm} being the Kronecker symbol ($\delta_{nm} = 1$ when $n = m$, $\delta_{nm} = 0$ when $n \neq m$). The quantities $\lambda_n(\omega)$ [the eigenvalues of the integral equation (5)] are positive, i.e.

$$\lambda_n(\omega) > 0 \quad (n \geq 0). \tag{7}$$

The summation in Eq. (4) must be interpreted as follows. If D is a three-dimensional domain, n stands for the triplet (n_1, n_2, n_3) of non-negative integers and Σ stands for a triple sum. If the domain is two-dimensional n stands for a pair of non-negative integers, n_1 and

n_2, and Σ stands for a double sum. If the domain is one-dimensional, n stands for a non-negative integer and one has a single summation.

In free space, each function $\phi_n(\mathbf{r}, \omega)$ obeys the Helmholtz equation

$$\nabla^2 \phi_n(\mathbf{r}, \omega) + k^2 \phi_n(\mathbf{r}, \omega) = 0. \tag{8}$$

To derive this result we substitute the expansion (4) of the cross-spectral density into the Helmholtz equation (3b), multiply the resulting equation by $\phi_m(\mathbf{r}_1, \omega)$, integrate both sides with respect to $\mathbf{r}_1$ over the domain D and use the orthonormality relation (6). For reasons which will become evident later, the expansion (4) is known as *the coherent-mode representation* of the cross-spectral density.

We will now show that by using the expansion (4) of the cross-spectral density, one can construct an ensemble $\{U(\mathbf{r}, \omega)\}$ of sample functions $U(\mathbf{r}, \omega)$ in terms of which the cross-spectral density of the field in the domain D may be expressed as a correlation function.

Let us consider the ensemble of sample functions of the form

$$U(\mathbf{r}, \omega) = \sum_n a_n(\omega) \phi_n(\mathbf{r}, \omega), \tag{9}$$

where the $a_n(\omega)$ are random coefficients such that

$$\langle a_n^*(\omega) a_m(\omega) \rangle_\omega = \lambda_n(\omega) \delta_{nm} \tag{10}$$

and the $\lambda_n(\omega)$ are the same positive quantities as appear in the expansion (4),[1] i.e. the eigenvalues of the integral equation (5).

Next we consider the correlation function $\langle U^*(\mathbf{r}_1, \omega) U(\mathbf{r}_2, \omega) \rangle$. One has, on using the expansion (9),

$$\langle U^*(\mathbf{r}_1, \omega) U(\mathbf{r}_2, \omega) \rangle_\omega = \sum_n \sum_m \langle a_n^*(\omega) a_m(\omega) \rangle \phi_n^*(\mathbf{r}_1, \omega) \phi_m(\mathbf{r}_2, \omega), \tag{11}$$

where we have interchanged the order of the ensemble average and the double summation. On using (10), the expansion (11) simplifies:

$$\langle U^*(\mathbf{r}_1, \omega) U(\mathbf{r}_2, \omega) \rangle_\omega = \sum_n \lambda_n(\omega) \phi_n^*(\mathbf{r}_1, \omega) \phi_n(\mathbf{r}_2, \omega). \tag{12}$$

Since the right-hand sides of Eqs. (12) and (4) are the same, the left-hand sides must be equal to each other and hence we have established the important result that

$$W(\mathbf{r}_1, \mathbf{r}_2, \omega) = \langle U^*(\mathbf{r}_1, \omega) U(\mathbf{r}_2, \omega) \rangle_\omega. \tag{13}$$

---

[1] There are many ways of choosing such random coefficients. For example one can take $a_n(\omega) = \sqrt{\lambda_n(\omega)} e^{i\theta_n}$, where, for each n, θ_n is a real random variable that is uniformly distributed in the range $0 \leqslant \theta_n < 2\pi$ and θ_n and θ_m are statistically independent when $n \neq m$. With this choice the requirement (10) is satisfied.

We have attached the suffix ω on the angular brackets to stress that the average is taken over an ensemble of space–frequency realizations. It is an ensemble which is quite different from the ensemble taken over the space–time realizations $V(\mathbf{r}, t)$, which we encountered earlier.

The formula (13) is an important result. It shows that the *cross-spectral density of a statistically stationary fluctuating field in the domain D may be expressed, for all pairs of points in D, as a cross-correlation function of an ensemble* $\{U(\mathbf{r}, \omega)\}$ *of space–frequency realizations* $U(\mathbf{r}, \omega)$.

Because each of the functions $\phi_n(\mathbf{r}, \omega)$ satisfies the Helmholtz equation (8) it is clear that the left-hand side of Eq. (9) also satisfies that equation and, consequently,

$$\nabla^2 U(\mathbf{r}, \omega) + k^2 U(\mathbf{r}, \omega) = 0. \tag{14}$$

Hence we may regard each sample function $U(\mathbf{r}, \omega)$ of our ensemble as being the space-dependent part of a monochromatic wavefield $V(\mathbf{r}, t) = U(\mathbf{r}, \omega)\exp(-i\omega t)$. This fact contributes towards an intuitive understanding of many of the results of the second-order coherence theory in the space–frequency domain. For example, Eq. (13) implies that the spectral density $S(\mathbf{r}, \omega) \equiv W(\mathbf{r}, \mathbf{r}, \omega)$ of the fluctuating field $V(\mathbf{r}, t)$ at a point $\mathbf{r}$ may be expressed in the form

$$S(\mathbf{r}, \omega) = \langle U^*(\mathbf{r}, \omega)U(\mathbf{r}, \omega)\rangle_\omega. \tag{15}$$

This formula is similar to the formula based on the common intuitive belief that the spectral density is the average of the squared modulus of the Fourier frequency components of the fluctuating field $V(\mathbf{r}, t)$. However, as we learned in Section 2.5, a stationary random field $V(\mathbf{r}, t)$ does not have a Fourier frequency spectrum. Nevertheless the formula (15) is rigorously valid, but one must appreciate that $U(\mathbf{r}, \omega)$ is *not* a Fourier frequency component (which does not exist) of the fluctuating field but is the space-dependent part of a member of the *statistical ensemble* $\{V(\mathbf{r}, t) = U(\mathbf{r}, \omega)e^{-i\omega t}\}$ *of monochromatic realizations*, all of frequency ω. The distinction between a monochromatic field and an ensemble of mono-chromatic fields of the same frequency is crucial. Once this fact is appreciated, one can use the space–frequency representation with great advantage to study second-order coherence phenomena in stationary wavefields, as we will soon see.

4.2 The spectral interference law and the spectral degree of coherence

In Section 3.1 we introduced a (generally complex) space–time correlation coefficient, namely the degree of coherence $\gamma(\mathbf{r}_1, \mathbf{r}_2, \tau)$, from the analysis of Young's interference experiment. In this section we will introduce a space–frequency correlation coefficient, also from the analysis of Young's interference experiment, but with the difference that instead of considering the distribution of the intensity in the detection plane we will consider the spectrum of the light in that plane. For this purpose it is *not* necessary that the light illuminating the pinholes is narrow-band; on the contrary, the effect of superposing two beams emerging from the pinholes on the spectrum of the light in the detection plane

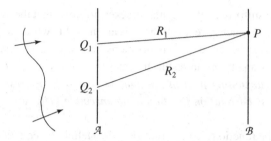

Fig. 4.1 Notation relating to Young's interference experiment with broad-band light.

becomes more pronounced when the light incident on the pinhole contains a broad band of frequencies, as we will soon see.

Let us once again consider light of any state of coherence, incident from the left on an opaque screen $\mathcal{A}$ containing two pinholes, at points Q_1 and Q_2 (Fig. 4.1). As is clear from the results that we have just established, we may represent the field at the pinholes by ensembles of frequency-dependent realizations $\{U(Q_1, \omega)\}$ and $\{U(Q_2, \omega)\}$. We assume that the pinholes are sufficiently small that the amplitude of the field is effectively constant over each of them and also that the angles of incidence and of diffraction are small. Then the field at a point P on a detection plane $\mathcal{B}$ some distance beyond the screen $\mathcal{A}$ and parallel to it is, to a good approximation, given by an ensemble of realizations $\{U(P, \omega)\}$ where

$$U(P, \omega) = K_1 U(Q_1, \omega) e^{ikR_1} + K_2 U(Q_2, \omega) e^{ikR_2}, \tag{1}$$

K_1 and K_2 being defined by Eq. (3) of Section 3.1, with $\overline{\lambda}$ being replaced by the wavelength λ corresponding to the frequency ω, i.e. $\lambda = 2\pi c/\omega$ and R_1 and R_2 are, as before, the distances from Q_1 to P and from Q_2 to P, respectively.

Let us substitute from Eq. (1) into the expression for the spectral density [Eq. (15) of Section 4.1]. If we also use the fact that $W(Q_2, Q_1, \omega) = W^*(Q_1, Q_2, \omega)$, we obtain for the spectral density at the point P the expression

$$S(P, \omega) = |K_1|^2 S(Q_1, \omega) + |K_2|^2 S(Q_2, \omega) + 2\,\mathcal{R}e\{K_1^* K_2 W(Q_1, Q_2, \omega) e^{-i\delta}\}, \tag{2}$$

where

$$\delta = \frac{2\pi}{\lambda}(R_1 - R_2). \tag{3}$$

The factors K_1 and K_2 are proportional to the areas of the pinholes. If we let the area of the pinhole at Q_2 decrease to zero, Eq. (2) will then represent the spectral density, $S^{(1)}(P, \omega)$ say, at the point P, with just the pinhole at Q_1 open, i.e.

$$|K_1|^2 S(Q_1, \omega) \equiv S^{(1)}(P, \omega). \tag{4a}$$

Similarly,

$$|K_2|^2 S(Q_2, \omega) \equiv S^{(2)}(P, \omega). \tag{4b}$$

represents the spectral density at P with just the pinhole at Q_2 open. Hence the formula (2) may be rewritten in the physically more significant form

$$S(P, \omega) = S^{(1)}(P, \omega) + S^{(2)}(P, \omega)$$
$$+ 2\left\{ \sqrt{S^{(1)}(P, \omega)} \sqrt{S^{(2)}(P, \omega)} \, \mathcal{R}e[\mu(Q_1, Q_2, \omega)e^{-i\delta}] \right\}, \tag{5}$$

where

$$\mu(Q_1, Q_2, \omega) \equiv \frac{W(Q_1, Q_2, \omega)}{\sqrt{W(Q_1, Q_1, \omega)} \sqrt{W(Q_2, Q_2, \omega)}} \tag{6a}$$

$$= \frac{W(Q_1, Q_2, \omega)}{\sqrt{S(Q_1, \omega)} \sqrt{S(Q_2, \omega)}}. \tag{6b}$$

If we set

$$\mu(Q_1, Q_2, \omega) = |\mu(Q_1, Q_2, \omega)| e^{i\beta(Q_1, Q_2, \omega)} \tag{7}$$

the expression (5) for the spectral density at the point P in the observation plane becomes

$$S(P, \omega) = S^{(1)}(P, \omega) + S^{(2)}(P, \omega)$$
$$+ 2\sqrt{S^{(1)}(P, \omega)} \sqrt{S^{(2)}(P, \omega)} |\mu(Q_1, Q_1, \omega)| \cos[\beta(Q_1, Q_2, \omega) - \delta]. \tag{8}$$

This formula is called the *spectral interference law* for the superposition of beams of any state of coherence. In its mathematical structure it is of the same form as the "intensity interference law" [Eq. (14) of Section 3.1]; but its meaning is different. The intensity interference law is an expression for the averaged *intensity* at a point P in the interference pattern, whereas the formula (8) is an expression for the *spectral density* of the light at that point.

Before turning our attention to some consequences of the spectral interference law, we will briefly comment on the physical significance of the factor $\mu(Q_1, Q_2, \omega)$ which plays an analogous role to the complex degree of coherence $\gamma(Q_1, Q_2, \tau)$. This is evident on comparing Eq. (8) with Eq. (14) of Section 3.1. The expressions (6) by which $\mu(Q_1, Q_2, \omega)$ has been introduced show that it is the normalized cross-spectral density of the fluctuating field at the points Q_1 and Q_2. We have already learned that the cross-spectral density $W(\mathbf{r}_1, \mathbf{r}_2, \omega)$ may be interpreted as a correlation function in the space–frequency domain [Eq. (12) of Section 4.1]. By using the fact that the cross-spectral density function is non-negative definite (or by use of the Schwarz inequality) one may show [see B&W, Appendix VIII, p. 911] that its normalized factor $\mu(Q_1, Q_2, \omega)$ is bounded in absolute value by zero and unity, i.e. that

$$0 \leq |\mu(Q_1, Q_2, \omega)| \leq 1 \tag{9}$$

for all values of its argument. The extreme value $|\mu| = 1$ represents complete correlation; the other extreme value, $\mu = 0$, represents absence of correlation. For this reason the normalized cross-spectral density $\mu(Q_1, Q_2, \omega)$ is called the *spectral degree of coherence* at frequency ω of the light at the points Q_1 and Q_2.

Let us now return to the spectral interference law (8). Usually $S^{(2)}(P, \omega) \approx S^{(1)}(P, \omega)$ and the spectral interference law then takes the simpler form

$$S(P, \omega) = 2S^{(1)}(P, \omega)\{1 + |\mu(Q_1, Q_2, \omega)|\cos[\beta(Q_1, Q_2, \omega) - \delta]\}. \qquad (10)$$

We note two implications of this formula. First, at any frequency ω, the spectral density varies sinusoidally with the position of the point P across the detection plane $\mathcal{B}$, with the amplitude and phase depending on the spectral degree of coherence. Secondly, at any fixed point P the spectral density $S(P, \omega)$ will, in general, differ from the spectral density $S^{(1)}(P, \omega)$ of the light which would reach the point P through only one of the two pinholes, the difference depending on the spectral degree of coherence $\mu(Q_1, Q_2, \omega)$. This difference in the two spectra is an example of the phenomenon of *correlation-induced spectral changes*, which we will discuss shortly.

In a sense the "intensity" interference law given by Eq. (8) of Section 3.1 and the spectral interference law (8) are complementary to each other. The former shows that appreciable modifications of the averaged *intensity* take place when *narrow-band* quasi-monochromatic light beams are superposed. The latter indicates that appreciable changes of *spectra* may take place when two *broad-band* beams are superposed. More detailed analysis reveals that in the former case no appreciable spectral changes take place, whereas in the latter case no appreciable intensity variations occur. Moreover, no interference fringes are formed when the path difference introduced between the two beams exceeds distances of the order of the coherence length of the light, whereas spectral modulation takes place irrespective of the phase difference[1] δ, defined by Eq. (3), as is evident from the spectral interference law (8). Figure 4.2(a) shows the experimental set-up for illustrating this effect. Results of the experiment are shown in Fig. 4.2(b).

Spectral changes have also been observed in star light passed through two slits in an opaque screen and then superposed.[2] From such changes, shown in Fig. 4.3, the spectral degree of coherence of the star light reaching the Earth was determined by the use of the spectral interference law (8). In principle, one can estimate from such measurements the angular diameter of the star with the help of the van Cittert–Zernike theorem (that we discussed in Section 3.2). This method must be distinguished from Michelson's method for measuring stellar diameters, which we described in Section 3.3.1. Michelson's method is based on determining the correlation between the fluctuating fields at two mirrors of an interferometer from visibility measurements, whereas the method which we just mentioned

---

[1] Similar effects have been found in interference experiments with matter waves, specifically with neutron beams [see, for example H. Rauch, *Phys. Lett.* **A173** (1992), 240–242 and D. L. Jacobson, S. A. Werner and H. Rauch, *Phys. Rev.* **A49** (1994), 3196–3200. See also G. S. Agarwal, *Found. Phys.* **25** (1995), 219–228.

[2] H. C. Kandpal, A. Wasan, J. C. Vaishya and E. S. R. Gopal *Indian J. Pure Appl. Phys.* **36** (1998), 665–674. In this paper measurements of the frequency dependence of the spectral degree of coherence of light from a star (α-Bootis) were also reported.

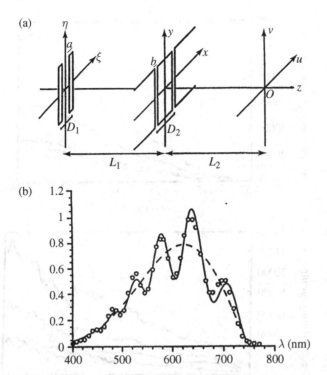

Fig. 4.2 Spectral changes generated on superposition of two partially coherent, broad-band light beams, emerging from two slits. (a) Layout of the experiment with $D_1 = 0.68\,\text{mm}$, $D_2 = 3.4\,\text{mm}$, $a = 0.026\,\text{mm}$, $b = 0.11\,\text{mm}$. (b) Measured values, denoted by circles and interpolated by the solid curves and the original spectrum (broken line). [After M. Santarsiero and F. Gori, *Phys. Lett.* **A167** (1992), 123–128.]

is based on measurements of the spectral changes arising on interference between two beams, i.e. it makes use of *correlation-induced* spectral changes.

In discussing the "intensity interference law" in Section 3.1, we showed that the absolute value of the complex degree of coherence $\gamma(Q_1, Q_2, \tau)$ can be determined from measurements of the visibility of interference fringes [Eq. (19) of Section 3.1]. We will now show that the absolute value of the spectral degree of coherence $\mu(Q_1, Q_2, \omega)$ may be determined in a somewhat similar manner. For this purpose we note from the spectral interference law, specialized to the case when $S^{(2)}(P, \omega) = S^{(1)}(P, \omega)$ [Eq. (10)], that the spectral density $S(P, \omega)$ at any fixed frequency ω has a maximum in the neighborhood of a point P in the plane of observation, when the path difference $(\beta - \delta)$ is such that the cosine term in Eq. (10) has the value $+1$ and that it has a minimum when it has the value -1. These extreme values of the spectral density evidently are

$$S_{\text{max}}(P, \omega) = 2S^{(1)}(\omega)[1 + |\mu(Q_1, Q_2, \omega)|], \tag{11a}$$

$$S_{\text{min}}(P, \omega) = 2S^{(1)}(\omega)[1 - |\mu(Q_1, Q_2, \omega)|]. \tag{11b}$$

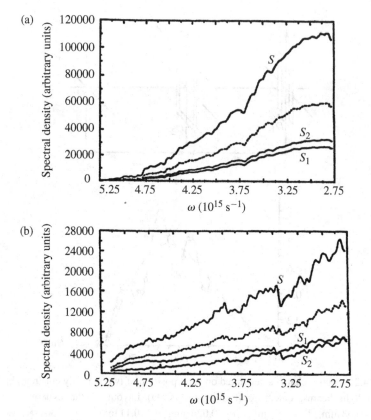

Fig. 4.3 Spectra of the stars α-Bootis (a) and α-Scorpio (b) at an observation point P on axis. Curve S_1 is the recorded spectrum when slit P_1 is open and P_2 is closed. The curve S_2 is the recorded spectrum when the slit P_2 is open and P_1 is closed. Curve S is the recorded spectrum when both P_1 and P_2 are open. The dotted line represents the sum of the two spectra given by the curves S_1 and S_2. [After H. C. Kandpal, A. Wasan, J. S. Vaishya and E. S. R. Gopal, *Indian J. Pure Appl. Phys.* **36** (1998), 665–674.]

By analogy with the definition Eq. (18) of Section 3.1 of the fringe visibility, we now introduce the concept of *spectral visibility* $\mathcal{V}(P, \omega)$, defined by the expression

$$\mathcal{V}(P,\omega) = \frac{S_{\max}(P,\omega) - S_{\min}(P,\omega)}{S_{\max}(P,\omega) + S_{\min}(P,\omega)}. \tag{12}$$

On substituting from Eqs. (11) into Eq. (12) it follows that

$$\mathcal{V}(P,\omega) = |\mu(Q_1, Q_2, \omega)|. \tag{13}$$

This formula shows that the absolute value of the spectral degree of coherence can be obtained from Young's interference experiment by letting the light from the two pinholes pass through narrow-band filters which transmit a narrow portion of the spectrum centered on the selected frequency ω. The argument (phase) of μ can also be determined experimentally, from measurements of the locations of the maxima and the minima, in a similar manner (mentioned in Section 3.1 and described in B&W, Section 10.4.1) to what can be done to determine the phase of γ; but, in addition, one must again use narrow-band filters. Other techniques for measuring both the modulus and the phase of the spectral degree of coherence have been described in several publications.[1]

With the interpretation of the normalized cross-spectral density (6) as the spectral degree of coherence, one can readily understand why the expansion Eq. (4) of Section 4.1 is known as the coherent-mode representation of the field. For this purpose let us rewrite the expansion in the form

$$W(\mathbf{r}_1, \mathbf{r}_2, \omega) = \sum_n W^{(n)}(\mathbf{r}_1, \mathbf{r}_2, \omega),\tag{14}$$

where

$$W^{(n)}(\mathbf{r}_1, \mathbf{r}_2, \omega) = \lambda_n(\omega)\phi_n^*(\mathbf{r}_1, \omega)\phi_n(\mathbf{r}_2, \omega).\tag{15}$$

The spectral degree of coherence associated with the contribution of $W^{(n)}$ is given by the expression

$$\mu^{(n)}(\mathbf{r}_1, \mathbf{r}_2, \omega) \equiv \frac{W^{(n)}(\mathbf{r}_1, \mathbf{r}_2, \omega)}{\sqrt{S^{(n)}(\mathbf{r}_1, \omega)}\sqrt{S^{(n)}(\mathbf{r}_2, \omega)}}$$

$$= \frac{\lambda_n(\omega)\phi_n^*(\mathbf{r}_1, \omega)\phi_n(\mathbf{r}_2, \omega)}{\sqrt{\lambda_n(\omega)|\phi_n(\mathbf{r}_1, \omega)|^2}\sqrt{\lambda_n(\omega)|\phi_n(\mathbf{r}_2, \omega)|^2}},\tag{16}$$

implying that

$$|\mu^{(n)}(\mathbf{r}_1, \mathbf{r}_2, \omega)| = 1.\tag{17}$$

In going from the first to the second expression on the right of Eq. (16) we used the fact that the spectral density $S^{(n)}(\mathbf{r}, \omega) \equiv W^{(n)}(\mathbf{r}, \mathbf{r}, \omega)$.

Equation (17) shows that each term (mode) in the expansion Eq. (4) of Section 4.1 represents a field which is completely spatially coherent at frequency ω.

It may be shown (M&W, Section 7.4) that laser modes provide examples of such coherent modes.

4.3 An illustrative example: spectral changes on interferences

The spectral interference law that we have derived in the preceding section has a number of interesting implications and some useful applications. In this section we will show how it may be used to determine the angular separation of distant objects.[2]

---

[1] D. F. V. James and E. Wolf, *Opt. Commun.* **145** (1997), 1–4; S. S. K. Titus, A. Wasan, J. S. Vaishya and H. C. Kandpal, *Opt. Commun.* **173** (2000), 45–49; V. N. Kumar and D. N. Rao, *J. Mod. Opt.* **48** (2001), 1455–1465; G. Popescu and A. Dogariu, *Phys. Rev. Lett.* **88** (2002), 183902 (4 pages).
[2] The analysis of this section is based on a paper by D. F. V. James, H. C. Kandpal and E. Wolf, *Astrophys. J.* **445** (1995), Part I, 406–410.

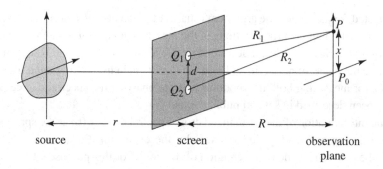

Fig. 4.4 Illustrating the notation relating to formula (2) of Section 4.3.

When the spectral densities of the light reaching the observation point P from each of the pinholes are the same [i.e. $S^{(1)}(P, \omega) = S^{(2)}(P, \omega)$] we have, according to the spectral interference laws [Eq. (10) of Section 5.2],

$$\frac{S(P,\omega)}{2S^{(1)}(P,\omega)} = 1 + |\mu(Q_1, Q_2, \omega)| \cos[\beta(Q_1, Q_2, \omega) - \delta]. \tag{1}$$

Let R_1 and R_2 denote, as before, the distances from each of the two pinholes at points Q_1 and Q_2 to the point of observation P, d the distance between the pinholes and x the distance of the point of observation from the axis (see Fig. 4.4). Suppose that x is small relative to the distance R between the plane of the pinholes and the observation plane, i.e. that $x/R \ll 1$. Then $R_2 - R_1 \approx xd/R$ and the path difference $\delta \equiv k(R_2 - R_1) \approx \omega xd/(Rc)$. Under these circumstances the formula (1) becomes

$$\frac{S(P,\omega)}{2S^{(1)}(P,\omega)} = 1 + |\mu(Q_1, Q_2, \omega)| \cos[\beta(Q_1, Q_2, \omega) - \omega xd/(Rc)]. \tag{2}$$

Suppose that the source is a pair of identical circular disks, e.g. a somewhat idealized double star, each of uniform intensity i_0. The intensity distribution $I_0(\rho)$ across this two-component source, each component of which is assumed to be spatially incoherent, may be expressed in the form

$$I_0(\rho) = i_0\{\mathrm{circ}[|\rho + \mathbf{b}_0/2|/a] + \mathrm{circ}[|\rho - \mathbf{b}_0/2|/a]\}, \tag{3}$$

where a denotes the radius of each of the two circular sources, $\mathbf{b}_0$ is the vector specifying the location of the center of one of them relative to the other, i_0 is the intensity, assumed to be constant, of either of the two sources and

$$\mathrm{circ}(x) = \begin{cases} 1 & \text{when } 0 \leq |x| \leq 1, \\ 0 & \text{when } |x| > 1. \end{cases} \tag{4}$$

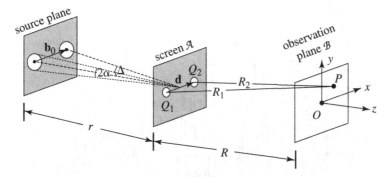

Fig. 4.5 Illustrating a method for determining the angular diameters and angular separation of two circular sources from spectroscopic measurements.

The spectral degree of coherence of the light reaching the pinholes from the source can readily be calculated by substituting from Eq. (3) into the far zone form Eq. (20) of Section 3.2, of the van Cittert–Zernike theorem with the equal-time degree of coherence $j(Q_1, Q_2)$ replaced by the spectral degree of coherence $\mu(Q_1, Q_2, \omega)$. One then finds that

$$\mu(Q_1, Q_2, \omega) = \frac{2J_1[a\omega d/(cr)]}{a\omega d/(cr)} \cos\left(\frac{\omega \mathbf{d} \cdot \mathbf{b}_0}{2cr}\right),\tag{5}$$

where r is the distance between the "source plane" and the plane of the pinholes (see Fig. 4.5) and $\mathbf{d}$ is the (vectorial) distance between the two pinholes. For the sake of simplicity we have assumed that the line joining the pinholes is parallel to the line joining the centers of the two circular sources, i.e. that $\mathbf{d}$ is parallel to $\mathbf{b}_0$.

On substituting from Eq. (5) into Eq. (2) it follows that

$$\frac{S(P, \omega)}{2S^{(1)}(P, \omega)} = 1 + \frac{2J_1[a\omega d/(cr)]}{a\omega/(cr)} \cos\left(\frac{\omega \mathbf{d} \cdot \mathbf{b}_0}{2cr}\right) \cos\left(\frac{\omega x d}{cR}\right),\tag{6}$$

where we have used the fact that the factor containing the Bessel function is real and positive and that $R \gg d$.

Figure 4.6 shows the behavior of the ratio $2S(P, \omega)/S^{(1)}(P, \omega)$ as a function of the frequency ω, for some selected values of the parameters. Several features are worth noting: first, the rapid sinusoidal modulation, due to the factor $\cos[\omega x d/(cR)]$; and secondly, the contrast of the spectral modulation due to the frequency dependence of the spectral degree of coherence, given by Eq. (5), of the light at the two pinholes. One can readily identify the two different causes of the modulation as being due to (i) the size of the source and (ii) the separation of the two sources. Thus, in principle, the angular radii of the sources and their angular separation can be deduced from such spectral measurements made at *fixed* pinhole separation. These theoretical predictions have been verified by experiments[1] (see Fig. 4.7).

[1] H. C. Kandpal, K. Saxena, D. S. Mehta, J. S. Vaishya and K. C. Joshi, *J. Mod. Opt.* **42** (1995), 447–454.

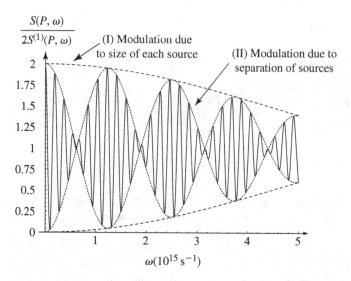

Fig. 4.6 The spectrum produced on superposing, at a point P, two beams from circular sources, each of angular radius $\alpha = 3 \times 10^{-8}$. The angular separation of the two sources is $\Delta = 3 \times 10^{-7}$, the path difference $(R_2 - R_1) = 10\,\mu\text{m}$ and the baseline $d = 5\,\text{m}$. [Adapted from D. F. V. James, H. C. Kandpal and E. Wolf, *Astrophys. J.* **445** (1995), 406–410.]

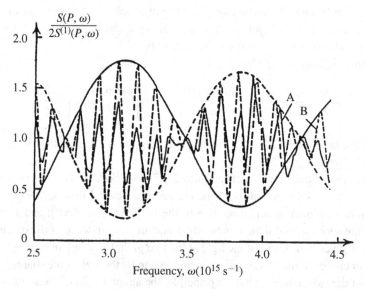

Fig. 4.7 Results of the first laboratory experiments testing the theoretical predictions shown in the previous figure, with the same values of the parameters. Dashed line: theoretical prediction; solid line: results of experiments. [Adapted from H. C. Kandpal, K. Saxena, D. S. Mehta, J. S. Vaishya and K. C. Joshi, *J. Mod. Opt.* **42** (1995), 447–454.]

4.4 Interference of narrow-band light

It is frequently assumed that, when light of sufficiently narrow bandwidth, centered at frequency ω_0, say, interferes, sharp fringes (i.e. fringes essentially of unit visibility) are formed, imitating interference of monochromatic light of that frequency. We will now show that this assumption is incorrect; and also that interference fringes formed with narrow-band light exhibit a number of interesting features.

Suppose that we place identical filters in front of each of the pinholes at points Q_1 and Q_2 in Young's experiment. We will consider how the passband of the filters affects the interference pattern.

Let

$$W^{(i)}(Q_1, Q_2, \omega) = \langle U^{(i)*}(Q_1, \omega)U^{(i)}(Q_2, \omega)\rangle \tag{1}$$

be the cross-spectral density of the light incident on the pinholes. The expectation value on the right of Eq. (1) is to be understood in the sense of coherence theory in the space-frequency domain, which we discussed in Section 4.1. For the sake of simplicity, we have now omitted the subscript ω on the angular brackets.

Let $T(\omega)$ be the transmission function of each filter. The cross-spectral density function of the filtered light which emerges from the filters is then given by the expression

$$W^{(+)}(Q_1, Q_2, \omega) = \langle T^*(\omega)U^{(i)*}(Q_1, \omega)T(\omega)U^{(i)}(Q_2, \omega)\rangle$$
$$= |T(\omega)|^2 W^{(i)}(Q_1, Q_2, \omega), \tag{2}$$

where we have made use of Eq. (1).

The spectral degree of coherence of the light transmitted by the filters, immediately behind them, is given by the expression

$$\mu^{(+)}(Q_1, Q_2, \omega) = \frac{W^{(+)}(Q_1, Q_2, \omega)}{\sqrt{W^{(+)}(Q_1, Q_1, \omega)}\sqrt{W^{(+)}(Q_2, Q_2, \omega)}}. \tag{3}$$

On substituting from Eq. (2) into Eq. (3) we find at once that

$$\mu^{(+)}(Q_1, Q_2, \omega) = \mu^{(i)}(Q_1, Q_2, \omega), \tag{4}$$

where

$$\mu^{(i)}(Q_1, Q_2, \omega) = \frac{W^{(i)}(Q_1, Q_2, \omega)}{\sqrt{W^{(i)}(Q_1, Q_1, \omega)}\sqrt{W^{(i)}(Q_2, Q_2, \omega)}}, \tag{5}$$

is the spectral degree of coherence of the light incident on the pinholes. Formula (4) shows that the *spectral degree of coherence is unchanged by linear filtering*.

Next let us consider the effect of the filters on the "space–time" degree of coherence $\gamma(Q_1, Q_2, \tau)$ which we discussed in Section 3.1. For light incident on the two pinholes, it is given by the expression

$$\gamma^{(i)}(Q_1, Q_2, \tau) = \frac{\Gamma^{(i)}(Q_1, Q_2, \tau)}{\sqrt{\Gamma^{(i)}(Q_1, Q_1, \tau)}\sqrt{\Gamma^{(i)}(Q_2, Q_2, \tau)}}, \tag{6}$$

where $\Gamma^{(i)}(Q_1, Q_2, \tau)$ is the mutual coherence function, defined by Eq. (1) of Section 4.1. According to the inverse of Eq. (2) of Section 4.1, it is just the Fourier transform of the cross-spectral density function, i.e.

$$\Gamma^{(i)}(Q_1, Q_2, \tau) = \int_0^\infty W^{(i)}(Q_1, Q_2, \omega)e^{-i\omega\tau}\, d\omega, \tag{7}$$

where the integration extends over the positive frequencies only, because of our use of the analytic signal representation.

On using Eq. (2) it is clear that the mutual coherence function of the light emerging from the filters is

$$\Gamma^{(+)}(Q_1, Q_2, \tau) = \int_0^\infty |T(\omega)|^2\, W^{(i)}(Q_1, Q_2, \omega)e^{-i\omega\tau}\, d\omega. \tag{8}$$

Suppose that the effective bandwidth $\Delta\omega$ of the filters is so small that the absolute value and the phase of the cross-spectral density $W^{(i)}(Q_1, Q_2, \omega)$ of the incident light are effectively constant over the bandwidth $\Delta\omega$ of the incident light (see Fig. 4.8). If ω_0 is the mean frequency of the incident light we may then evidently approximate Eq. (8) by the formula

$$\Gamma^{(+)}(Q_1, Q_2, \tau) \approx W^{(i)}(Q_1, Q_2, \omega_0)\int_0^\infty |T(\omega)|^2 e^{-i\omega\tau}\, d\tau. \tag{9}$$

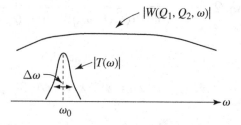

Fig. 4.8 Schematic illustration of the relative behavior of the modulus of the transmission function $T(\omega)$ of identical filters placed in front of the pinholes in Young's interference experiment and of the modulus of the cross-spectral density $W(Q_1, Q_2, \omega)$ of the light incident on the pinholes, in experiments with narrow-band light. The effective pass-band $\omega_0 - \Delta\omega/2 \leq \omega \leq \omega_0 + \Delta\omega/2$ of the filtered light is assumed to be so narrow that the modulus and the phase (not shown in the figure) are substantially constant across it. [Adapted from E. Wolf, *Opt. Lett.* **8** (1983), 250–252.]

Hence the degree of coherence of the light emerging from the filters is given by the expression

$$\gamma^{(+)}(Q_1, Q_2, \tau) \equiv \frac{\Gamma^{(+)}(Q_1, Q_2, \tau)}{\sqrt{\Gamma^{(+)}(Q_1, Q_1, 0)}\sqrt{\Gamma^{(+)}(Q_2, Q_2, 0)}}$$

$$= \mu^{(i)}(Q_1, Q_2, \omega_0)\Theta(\tau), \qquad (10)$$

where the spectral degree of coherence $\mu^{(i)}$ is given by Eq. (3) and

$$\Theta(\tau) = \frac{\int_0^\infty |T(\omega)|^2 e^{-i\omega\tau} \, d\omega}{\int_0^\infty |T(\omega)|^2 \, d\omega}. \qquad (11)$$

We will refer to the function $\Theta(\tau)$ as the filter function. We note that, as a consequence of a well-known inequality for integrals, we have

$$\underset{\tau}{\text{Max}}|\Theta(\tau)| = \Theta(0) = 1. \qquad (12)$$

It follows from Eq. (12) that

$$\underset{\tau}{\text{Max}}|\gamma^{(+)}(Q_1, Q_2, \tau)| = |\mu^{(i)}(Q_1, Q_2, \omega_0)|. \qquad (13)$$

In words: *the maximum of the absolute value of the (temporal) degree of coherence of the light emerging from the filters behind the two pinholes is equal to the absolute value of the spectral degree of coherence at the central frequency ω_0 of the light incident on the pinholes.*

Suppose that the averaged intensities of the light that is incident on the two pinholes are the same. Then, according to Eq. (19) of Section 3.1, the modulus of the degree of coherence $\gamma^{(+)}(Q_1, Q_2, \tau)$ is equal to the visibility of fringes formed by the light emerging from the two pinholes. Equation (13), therefore, implies that the maximum visibility of the fringes formed by the filtered light is equal to the modulus of the spectral degree of coherence $\mu^{(i)}(Q_1, Q_2, \omega_0)$ of the (unfiltered) light that is incident on the pinholes. Thus we see that *reducing the bandwidth of the incident light by linear filtering will not produce sharper fringes.* In particular the fringe visibility will not approach the value unity, irrespective of how narrow the passbands of the filters are (unless, of course, $|\mu^{(i)}(Q_1, Q_2, \omega_0)| = 1$). However, it is clear from Eq. (10) and from a well-known reciprocity inequality relating to effective widths of a pair of functions that are Fourier transforms of each other [B&W, Eq. (32) of Section 10.8] that the following result holds: the narrower the passbands of the filters, the broader the absolute value of the filter function $\Theta(\tau)$. Consequently, according to Eq. (10) and the relation (19) of Section 3.1 between $|\gamma^{(+)}(Q_1, Q_2, \tau)|$ and the visibility, more fringes will then be formed.

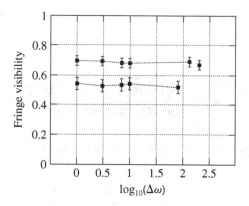

Fig. 4.9 Results of measurements of the fringe visibility $\mathcal{V}(\tau)$ in a two-pinhole experiment against $\log_{10}(\Delta\omega)$, where $\Delta\omega$ denotes the bandwidth of the filtered light. The two sets of measurements pertain to sets of filters centered on the wavelengths $\lambda = 633\,\text{nm}$ and $\lambda = 488\,\text{nm}$. The points on each sequence are joined by dotted lines for clarity. The measurements confirm that in the frequency ranges considered the fringe visibility is independent of the bandwidth of the light over very wide wavelength ranges. [Adapted from L. Basano, P. Ottonello, G. Rottingni and M. Vicari, *Appl. Opt.* **42** (2003), 6239–6244.]

We may summarize the main result that we derived in this section by saying that we have shown that linear filtering does not increase the spatial coherence of light (i.e. no sharper fringes are formed) but it increases its temporal coherence (i.e. more fringes will become visible). These theoretical predictions were verified experimentally.[1] Some of the experimental results are reproduced in Fig. 4.9.

PROBLEMS

4.1 It was shown in Section 4.1 that the cross-spectral density function of a statistically stationary field occupying a finite domain D may be expressed in the form

$$W(\mathbf{r}_1, \mathbf{r}_2, \omega) = \langle U^*(\mathbf{r}_1, \omega)U(\mathbf{r}_2, \omega)\rangle,$$

where

$$U(\mathbf{r}, \omega) = \sum a_n(\omega)\phi_n(\mathbf{r}, \omega).$$

The functions $\phi_n(\mathbf{r}, \omega)$ are the eigenfunctions of an integral equation whose kernel is the cross-spectral density function $W(\mathbf{r}_1, \mathbf{r}_2, \omega)$. The $a_n(\omega)$ are random coefficients which satisfy the requirement that $\langle a_n^* a_m\rangle = \lambda_n\delta_{nm}$, δ_{nm} being the Kronecker symbol.

[1] L. Basano, P. Ottonello, G. Rottigni and M. Vicari, *Appl. Opt.* **42** (2003), 6239–6244.

Show that

(1) $\langle U(\mathbf{r}_1, \omega)U(\mathbf{r}_2, \omega)\rangle = 0$; and

(2) if $U^{(r)}$ and $U^{(i)}$ are the real and the imaginary parts of U then

$$\langle U^{(r)}(\mathbf{r}_1, \omega)U^{(r)}(\mathbf{r}_2, \omega)\rangle = \langle U^{(i)}(\mathbf{r}_1, \omega)U^{(i)}(\mathbf{r}_2, \omega)\rangle$$

and

$$\langle U^{(r)}(\mathbf{r}_1, \omega)U^{(i)}(\mathbf{r}_2, \omega)\rangle = -\langle U^{(i)}(\mathbf{r}_1, \omega)U^{(r)}(\mathbf{r}_2, \omega)\rangle.$$

Find also expressions for the real and for the imaginary parts of the cross-spectral density W in terms of correlation functions involving the real and the imaginary parts of U.

4.2 (a) Derive a relation between the space–time degree of coherence $\gamma_{12}(\tau)$ and the spectral degree of coherence $\mu_{12}(\omega)$ of the field at two points P_1 and P_2.

(b) Suppose that the light is quasi-monochromatic and that the normalized spectra at the two points are equal to each other, i.e. that

$$s_1(\omega) = s_2(\omega);$$

and also that μ_{12} is independent of frequency over the narrow bandwidth $\Delta\omega$ of the light. How does the relationship between $\gamma_{12}(\tau)$ and $\mu_{12}(\omega)$ simplify in the case when $\tau \ll 2\pi/\Delta\omega$?

4.3 A cross-spectral density of a planar, secondary source has the factorized form

$$W(\mathbf{r}_1, \mathbf{r}_2, \omega) = F[(\mathbf{r}_1 + \mathbf{r}_2)/2, \omega]G(\mathbf{r}_2 - \mathbf{r}_1, \omega).$$

Show that, in order that $F(\mathbf{r}, \omega)$ represents the spectral density and $G(\mathbf{r}', \omega)$ the spectral degree of coherence of light across the source region, the spectral density has to satisfy a certain functional equation. Show that the above formula applies with any spectral density distribution whose spatial dependence has the form

$$S^{(0)}(\rho, \omega) \equiv S^{(0)}(x, y, \omega) = S^{(0)}(0, 0, \omega)e^{(\beta_1 x + \beta_2 y)},$$

where β_1 and β_2 are constants.

4.4 Consider two identical small sources, separated by a distance d. The spectrum of each source is $S_Q(\omega)$ and the correlation between them is characterized by $\mu_Q(\omega)$.

Derive an expression for the total power radiated by the two sources and discuss the limiting cases $d \ll \lambda$ and $d \gg \lambda$ ($\lambda = 2\pi c/\omega$). Comment on the implications of the result for the case $\lambda \gg d$ on the overall behavior of the spectrum of the far field.

4.5 Statistically stationary partially coherent light propagates from the input plane $z = z_0$ through a deterministic, time-invariant system, which is rotationally symmetric about the z-axis, to the output plane $z = z_1$. The system is characterized by an impulse response function $K(\rho, \rho', \omega)$. The vectors ρ and ρ' are two-dimensional position vectors perpendicular to the z-axis, of points in the input plane and in the output plane, respectively.

(a) If $W_0(\rho_1', \rho_2', \omega)$ is the cross-spectral density of the light in the input plane, derive an expression for the spectral density $S_1(\rho, \omega)$ of the light at a typical point in the output plane.

(b) Consider the special case of (a), when the input is a polychromatic plane wave which propagates in the positive-z direction.

4.6 $$Q_1(\omega) = \alpha(\omega)X(\omega) + Y(\omega), \qquad Q_2(\omega) = \beta(\omega)X(\omega) + Y(\omega)$$

are sample functions, in the frequency domain, that represent the fluctuations of two small sources, located at points P_1 and P_2. $X(\omega)$ and $Y(\omega)$ are random mutually uncorrelated functions, i.e.

$$\langle X^*(\omega)Y(\omega)\rangle = 0,$$

and $\alpha(\omega)$ and $\beta(\omega)$ are deterministic functions such that

$$|\alpha(\omega)| = |\beta(\omega)|.$$

(a) Derive an expression for the degree of correlation $\mu_{12}(\omega)$ of the two sources in terms of the spectra $S_X(\omega)$ and $S_Y(\omega)$ of $X(\omega)$ and $Y(\omega)$, respectively.

(b) Derive an expression for the spectrum $S_U(P, \omega)$ of field produced at a point P equidistant from the two sources and sufficiently far from them when

$$S_X(\omega) = S_Y(\omega) = A^2 e^{-(\omega-\omega_0)^2/(2\sigma^2)} \quad (A \text{ and } \sigma \text{ are positive constants})$$

and

$$|\alpha(\omega)| = 1, \qquad \phi_\alpha(\omega) - \phi_\beta(\omega) = 2\omega\tau, \quad (\tau \gg 1/\sigma),$$

where ϕ_α and ϕ_β are the phases of α and β respectively. Sketch $S_U(P, \omega)$ as a function of ω.

5

Radiation from sources of different states of coherence

5.1 Fields generated by sources with different coherence properties

Light generated by sources of different states of coherence may exhibit very different behavior. This is quite evident from some simple examples, as we will now show.

Suppose first that the light originates in a thermal source, e.g. it is emitted by a hot body. The radiant intensity, $J_\omega(\mathbf{s})$, which represents the rate at which power at frequency ω is radiated by the source per unit solid angle around a direction specified by a unit vector $\mathbf{s}$ (which makes an angle θ with the normal $\mathbf{n}$ to the source plane), is given by Lambert's law [Fig. 5.1(a)][1]

$$J_\omega(\theta) = J_\omega(0)\cos\theta. \tag{1}$$

This is a rather broad angular distribution, illustrated on a polar diagram in Fig. 5.1(a). On the other hand, light generated by a single-mode laser will be very directional [Fig. 5.1(b)]. Practically all the laser light will be concentrated within a very narrow solid angle around the forward direction and, consequently, the polar diagram of the radiant intensity $J_\omega(\mathbf{s})$ will now have a needle-like form.

Apart from this obvious difference between the radiation originating in these two kinds of sources there is a more subtle difference. It becomes evident if one considers the dependence of the radiant intensity on the shape of the source. It is clear from Eq. (1) that *the radiant intensity of the light generated by a Lambertian source is independent of the shape of the source, being just proportional to* cos θ; it is, therefore, rotationally symmetric about the normal $\theta = 0$ to the source plane. On the other hand, in view of the well-known Fourier-transform relationship between the light distribution in the far zone and the light distribution across the source plane of a spatially coherent source (usually the aperture plane), the radiant intensity generated by such a source depends strongly on the shape of the source. Consider, for example, a uniform circular source [Fig. 5.2(a)]. It will generate

---

[1] Examples of modern Lambertian sources, i.e. sources which obey Lambert's law, are light-emitting diodes. See, for example, E. F. Schumba, *Light-Emitting Diodes*, second edition (Cambridge University Press, Cambridge, 2006), Section 5.5.

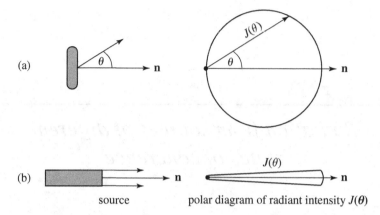

Fig. 5.1 Comparison of the angular distributions of the radiant intensity $J(\theta)$ of light produced by a thermal source (hot body) (a) and a single-mode laser (b).

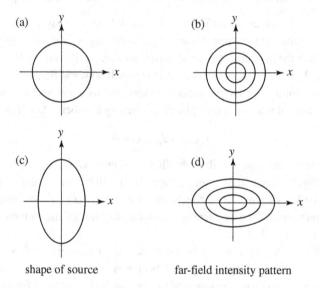

shape of source   far-field intensity pattern

Fig. 5.2 Illustrating the effect of change of shape of a spatially coherent planar source on the far-zone intensity pattern.

a rotationally symmetric intensity distribution, with circular contours [Fig. 5.2(b)]. Suppose that the source is "stretched" in the y direction [Fig. 5.2(c)]. Then the far-zone intensity distribution will shrink in that direction [Fig. 5.2(d)]. The difference between the two very different kinds of sources is evidently due to the difference in their coherence properties, the Lambertian source being spatially highly incoherent (see Section 5.5), whereas the laser is spatially highly coherent.

There are, of course, other important differences between beams of light generated by these two kinds of sources. As we have seen earlier (Section 2.1), the probability distributions which

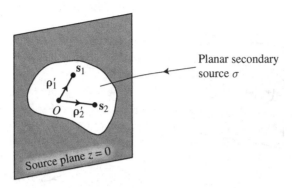

Fig. 5.3 Illustrating the notation relating to a planar, secondary source σ.

govern the fluctuations of the light field in the two cases are quite different. Another difference is in the average number of photons in a cell of phase space (the coherence volume). The number is called the *degeneracy parameter* of the light and is discussed in Appendix I.

In this chapter we will study radiation properties and coherence properties of fields generated by sources of different states of spatial coherence. We will mainly deal with fields generated by planar, secondary sources, which are of particular interest in many applications, e.g. in instrumental optics. The corresponding results pertaining to three-dimensional primary sources are very similar (M&W, Section 5.2).

5.2 Correlations and the spectral density in the far field

Let us consider a planar, secondary source σ of finite size, assumed to be statistically stationary, at least in the wide sense. Such a source may be an opening, for example, in an opaque screen illuminated either directly or via an optical system.

According to the coherence theory in the space–frequency domain that we studied in the preceding chapter, the cross-spectral density function of the field at a pair of points S_1 and S_2 in the source plane (see Fig. 5.3) may be expressed in the form [Eq. (13) of Section 4.1, with the subscript ω omitted from now on]

$$W^{(0)}(\rho_1', \rho_2', \omega) = \langle U^{(0)*}(\rho_1', \omega) U^{(0)}(\rho_2', \omega) \rangle. \tag{1}$$

Here ρ_1' and ρ_2' are two-dimensional position vectors specifying the locations of the two points, with respect to an origin O in the source plane $z = 0$, and the superscript zero indicates that the quantities pertain to points in that plane. $U^{(0)}(\rho', \omega)$ represents, of course, a member of a statistical ensemble of the frequency-dependent realizations and the angular brackets denote the average over that ensemble.

Let us now consider the cross-spectral density function $W(\mathbf{r}_1, \mathbf{r}_2, \omega)$ of the field at any pair of points $P_1(\mathbf{r}_1)$ and $P_2(\mathbf{r}_2)$ in the half-space $z > 0$ into which the source radiates (see Fig. 5.4). It may be expressed by the same formula as $W^{(0)}$, viz.,

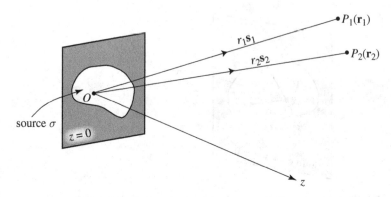

Fig. 5.4 Illustrating the notation relating to radiation from a planar, secondary source σ.

$$W(\mathbf{r}_1, \mathbf{r}_2, \omega) = \langle U^*(\mathbf{r}_1, \omega)U(\mathbf{r}_2, \omega)\rangle, \tag{2}$$

where $\{U(\mathbf{r}, \omega)\}$ represents the ensemble of the field at the point $\mathbf{r}$ generated by the field in the source plane $z = 0$. $U(\mathbf{r}, \omega)$ may be expressed in terms of the "boundary field" $U^{(0)}(\boldsymbol{\rho}', \omega)$ by the first Rayleigh diffraction integral [M&W, Eq. (3.2-78)][1]

$$U(\mathbf{r}, \omega) = -\frac{1}{2\pi} \int_\sigma U^{(0)}(\boldsymbol{\rho}', \omega) \left[\frac{\partial}{\partial z}\left(\frac{e^{ikR}}{R}\right)\right] d^2\rho', \tag{3}$$

where

$$R = |\mathbf{r} - \boldsymbol{\rho}'|. \tag{4}$$

We will consider the far field, i.e. the field at points P which are at very large distances from the origin, which is taken to be in the source region. To evaluate U at such points, it is convenient to set $\mathbf{r} = r\mathbf{s}$ ($\mathbf{s}^2 = 1$). Evidently, for sufficiently large distances r,

$$R \sim r - \mathbf{s} \cdot \boldsymbol{\rho}', \tag{5}$$

where $\mathbf{s} \cdot \boldsymbol{\rho}'$ denotes the projection ON of the distance OQ onto the $\mathbf{s}$ direction (see Fig. 5.5) and hence

---

[1] When the source is a primary three-dimensional source, rather than a two-dimensional secondary one, with source distribution $Q(\mathbf{r}, \omega)$ occupying a finite domain D, one has in place of Eq. (3) the formula

$$U(\mathbf{r}, \omega) = \int_D Q(\mathbf{r}', \omega) \frac{e^{ikR}}{R} d^3 r',$$

where $R = |\mathbf{r} - \mathbf{r}'|$. For proof of this result see, for example, C. H. Papas, *Theory of Electromagnetic Wave Propagation* (McGraw-Hill, New York, 1965), Section 2.1.

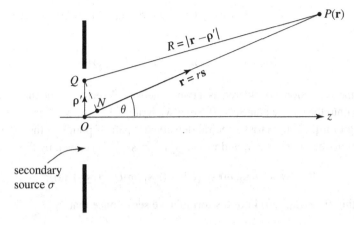

Fig. 5.5 Illustrating the approximation (5), valid for large distance r.

$$\frac{e^{ikR}}{R} \sim \frac{e^{ikr}}{r} e^{-iks\cdot\rho'}. \tag{6}$$

Consequently

$$\frac{\partial}{\partial z}\left(\frac{e^{ikR}}{R}\right) \sim ik\left(\frac{z}{r}\right)\frac{e^{ikr}}{r} e^{-iks\cdot\rho'}. \tag{7}$$

On substituting from Eq. (7) into the Rayleigh diffraction integral (3) and noting that $z/r = \cos\theta$ we obtain the following expression for the value of U in the far zone, which we will denote by $U^{(\infty)}$:

$$U^{(\infty)}(r\mathbf{s}, \omega) \sim -\frac{ik}{2\pi}\cos\theta \frac{e^{ikr}}{r}\int_{\sigma} U^{(0)}(\rho', \omega)e^{-iks\cdot\rho'}\,d^2\rho'. \tag{8}$$

Here θ denotes the angle which the $\mathbf{s}$ direction, pointing from the origin in the source plane to the observation point $P \equiv r\mathbf{s}$ in the far zone, makes with the z-axis (i.e. with the normal to the source plane).

It will be convenient to introduce the two-dimensional Fourier transform $\tilde{U}^{(0)}(\mathbf{f}, \omega)$ of $U^{(0)}(\rho', \omega)$, viz.,

$$\tilde{U}^{(0)}(\mathbf{f}, \omega) = \frac{1}{(2\pi)^2}\int\limits_{(z=0)} U^{(0)}(\rho', \omega)e^{-i\mathbf{f}\cdot\rho'}\,d^2\rho', \tag{9}$$

where $\mathbf{f}$ is a two-dimensional spatial-frequency vector. Although formally the integral on the right-hand side of Eq. (9) extends over the whole plane $z = 0$ containing the secondary source, it is actually taken only over the source region σ, because $U^{(0)}(\rho', \omega) = 0$ outside σ.

With the definition of $\tilde{U}^{(0)}$ given by Eq. (9), the formula (8) may be rewritten in the more compact form

$$U^{(\infty)}(r\mathbf{s}, \omega) \sim -2\pi i k \cos\theta \, \tilde{U}^{(0)}(k\mathbf{s}_\perp, \omega) \frac{e^{ikr}}{r}, \tag{10}$$

where $\mathbf{s}_\perp$ is the projection, considered as a two-dimensional vector, of the three-dimensional unit vector $\mathbf{s}$ onto the source plane $z = 0$; i.e. if we write $\mathbf{s} \equiv (s_x, s_y, s_z)$ then $\mathbf{s}_\perp \equiv (s_x, s_y, 0)$.

According to Eq. (2) the cross-spectral density at a pair of points in the far zone specified by position vectors $\mathbf{r}_1 \equiv r_1\mathbf{s}_1$ and $\mathbf{r}_2 \equiv r_2\mathbf{s}_2$ ($s_1^2 = s_2^2 = 1$) is given by the formula

$$W^{(\infty)}(r_1\mathbf{s}_1, r_2\mathbf{s}_2, \omega) = \langle U^{(\infty)*}(r_1\mathbf{s}_1, \omega)U^{(\infty)}(r_2\mathbf{s}_2), \omega)\rangle. \tag{11}$$

On substituting from Eq. (10) into this formula we see at once that

$$W^{(\infty)}(r_1\mathbf{s}_1, r_2\mathbf{s}_2, \omega) = (2\pi k)^2 \cos\theta_1 \cos\theta_2 \langle \tilde{U}^{(0)*}(k\mathbf{s}_{1\perp}, \omega) \tilde{U}^{(0)}(k\mathbf{s}_{2\perp}, \omega)\rangle \frac{e^{ik(r_2-r_1)}}{r_1 r_2}, \tag{12}$$

where, of course, θ_1 and θ_2 are the angles which the unit vectors $\mathbf{s}_1$ and $\mathbf{s}_2$ make with the positive z-axis. On the right of the formula (12) there is the average of the product of the two-dimensional Fourier components of $U^{(0)}$. We will now show that this average may be expressed as a four-dimensional Fourier transform of $W^{(0)}$. We have, on using the definition (9) of $\tilde{U}^{(0)}$, that

$$\langle \tilde{U}^{(0)*}(\mathbf{f}_1, \omega)\tilde{U}^{(0)}(\mathbf{f}_2, \omega)\rangle$$
$$= \frac{1}{(2\pi)^4} \iint_{(z=0)} \langle U^{(0)*}(\boldsymbol{\rho}_1', \omega)U^{(0)}(\boldsymbol{\rho}_2', \omega)\rangle e^{-i(\mathbf{f}_2\cdot\boldsymbol{\rho}_2' - \mathbf{f}_1\cdot\boldsymbol{\rho}_1')} d^2\rho_1' \, d^2\rho_2'. \tag{13}$$

Let us introduce the four-dimensional Fourier transform of $W^{(0)}$, viz.,

$$\tilde{W}^{(0)}(\mathbf{f}_1, \mathbf{f}_2, \omega) = \frac{1}{(2\pi)^4} \iint_{(z=0)} W^{(0)}(\boldsymbol{\rho}_1', \boldsymbol{\rho}_2', \omega) e^{-i(\mathbf{f}_1\cdot\boldsymbol{\rho}_1' + \mathbf{f}_2\cdot\boldsymbol{\rho}_2')} d^2\rho_1' \, d^2\rho_2'. \tag{14}$$

Now according to Eq. (1) the expectation value under the integral sign in Eq. (13) is just the cross-spectral density $W^{(0)}(\boldsymbol{\rho}_1', \boldsymbol{\rho}_2', \omega)$ of the field in the source plane and hence the right-hand sides of (13) and (14) are equal to each other, apart from the difference in the sign of $\mathbf{f}_1$ in the exponent. Taking this difference into account, the left-hand sides of these equations will also be equal to each other and hence

$$\langle \tilde{U}^{(0)*}(\mathbf{f}_1, \omega)\tilde{U}^{(0)}(\mathbf{f}_2, \omega)\rangle = \tilde{W}^{(0)}(-\mathbf{f}_1, \mathbf{f}_2, \omega). \tag{15}$$

Finally, on substituting from Eq. (15) into Eq. (12) we obtain the following expression for the cross-spectral density of the far field:

$$W^{(\infty)}(r_1\mathbf{s}_1, r_2\mathbf{s}_2, \omega) = (2\pi k)^2 \tilde{W}^{(0)}(-k\mathbf{s}_{1\perp}, k\mathbf{s}_{2\perp}, \omega) \frac{e^{ik(r_2-r_1)}}{r_1 r_2} \cos\theta_1 \cos\theta_2. \tag{16}$$

This is a basic formula from which various properties of the far field generated by planar, statistically stationary sources of any state of coherence may be deduced. We will study them later in this chapter.

The formula (16) is an analogue in second-order coherence theory of the Fraunhofer formula of the elementary theory of diffraction of monochromatic light. It implies that, apart from simple geometrical factors, the correlation of the radiated field at a pair of points in the far zone, in directions specified by unit vectors $\mathbf{s}_1$ and $\mathbf{s}_2$, is given by a particular four-dimensional spatial-frequency component $\tilde{W}^{(0)}$ $(\mathbf{f}_1, \mathbf{f}_2, \omega)$ of the cross-spectral density of the light in the source plane, namely one for which the two-dimensional spatial-frequency vectors are $\mathbf{f}_1 = -k\mathbf{s}_{1\perp}$ and $\mathbf{f}_2 = k\mathbf{s}_{2\perp}$. Since $\mathbf{s}_{1\perp}$ and $\mathbf{s}_{2\perp}$ are components of unit vectors it can be seen at once from Eq. (16) that only those spatial frequency vectors for which $|\mathbf{f}_1| \le k$ and $|\mathbf{f}_2| \le k$ of $W^{(0)}$ contribute to the cross-spectral density function $W^{(\infty)}$ of the far field. We will call them *low-spatial-frequency components*.

The spatial-frequency component for which the opposite inequalities hold, i.e. for which $|\mathbf{f}_1| > k$ and $|\mathbf{f}_2| > k$ may be called *high-spatial-frequency components*. There are, of course, also "mixed" situations when $|\mathbf{f}_1| > k$ and $|\mathbf{f}_2| \le k$ or $|\mathbf{f}_1| \le k$ and $|\mathbf{f}_2| > k$. The basic formula (16) shows that only the low-spatial-frequency components of the source contribute to the far field. The high-spatial-frequency components give rise to evanescent waves. (see, for example, M&W, Section 3.2), which decay exponentially in amplitude with increasing distance from the origin and, consequently, do not contribute to the far field.

As already mentioned, one may derive from formula (16) various properties of the far field. In particular let us consider the spectral density $S^{(\infty)}$ $(r\mathbf{s}, \omega)$, which is also called the optical intensity at frequency ω, of the field at the point $P(r\mathbf{s})$ in the far zone. It is given by the expression

$$S^{(\infty)}(r\mathbf{s}, \omega) = \langle U^{(\infty)*}(r\mathbf{s}, \omega)U^{(\infty)}(r\mathbf{s}, \omega)\rangle \equiv W^{(\infty)}(r\mathbf{s}, r\mathbf{s}, \omega) \tag{17}$$

or, using the formula (16),

$$S^{(\infty)}(r\mathbf{s}, \omega) = \left(\frac{2\pi k}{r}\right)^2 \tilde{W}^{(0)}(-k\mathbf{s}_\perp, k\mathbf{s}_\perp, \omega)\cos^2\theta. \tag{18}$$

We note the inverse-square-law dependence of $S^{(\infty)}$ on the distance r from the source, a result reminiscent of the inverse square-law of elementary wave theory. It is to be noted that the angular dependence of the spectral density is not only given by the proportionality factor $\cos^2\theta$ but also depends on the spatial coherence properties of the source, through the dependence of $\tilde{W}^{(0)}$ on the directional vector $\mathbf{s}$.

Because the distance from the source enters (18) only through the factor $1/r^2$ it is convenient to set

$$S^{(\infty)}(r\mathbf{s}, \omega) = \frac{J_\omega(\mathbf{s})}{r^2}, \tag{19}$$

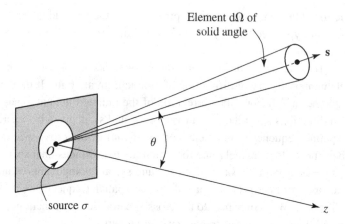

Fig. 5.6 Illustrating the meaning of the radiant intensity $J_\omega(\mathbf{s})$. It represents the rate at which energy at frequency ω is radiated into the far zone, per unit solid angle $d\Omega$ around the direction specified by a unit vector $\mathbf{s}$.

where, evidently,

$$J_\omega(\mathbf{s}) = (2\pi k)^2 \tilde{W}^{(0)}(-k\mathbf{s}_\perp, k\mathbf{s}_\perp, \omega)\cos^2\theta. \tag{20}$$

The function $J_\omega(\mathbf{s})$ is known as the *radiant intensity* at frequency ω or, more precisely, as the *spectral radiant intensity* and is a generalization of a quantity bearing the same name in traditional radiometry which deals with light from spatially incoherent sources. In suitable units the radiant intensity is a measure of the power radiated by the source per unit solid angle, around the direction specified by the unit vector $\mathbf{s}$, per unit frequency interval centered at the frequency ω (see Fig. 5.6).

Another quantity of interest relating to the far field is its spectral degree of coherence $\mu^{(\infty)}(r_1\mathbf{s}_1, r_2\mathbf{s}_2, \omega)$. According to the general formula (6a) of Section 4.2 it is given by the expression

$$\mu^{(\infty)}(r_1\mathbf{s}_1, r_2\mathbf{s}_2, \omega) = \frac{W^{(\infty)}(r_1\mathbf{s}_1, r_2\mathbf{s}_2, \omega)}{\sqrt{W^{(\infty)}(r_1\mathbf{s}_1, r_1\mathbf{s}_1, \omega)}\sqrt{W^{(\infty)}(r_2\mathbf{s}_2, r_2\mathbf{s}_2, \omega)}}. \tag{21}$$

Apart from a simple phase factor, the right-hand side of this formula may be expressed in terms of the cross-spectral density function of the source by the use of formula (16). One then finds that

$$\mu^{(\infty)}(r_1\mathbf{s}_1, r_2\mathbf{s}_2, \omega) = \frac{\tilde{W}^{(0)}(-k\mathbf{s}_{1\perp}, k\mathbf{s}_{2\perp}, \omega)}{\sqrt{\tilde{W}^{(0)}(-k\mathbf{s}_{1\perp}, k\mathbf{s}_{1\perp}, \omega)}\sqrt{\tilde{W}^{(0)}(-k\mathbf{s}_{2\perp}, k\mathbf{s}_{2\perp}, \omega)}} e^{ik(r_2-r_1)}. \tag{22}$$

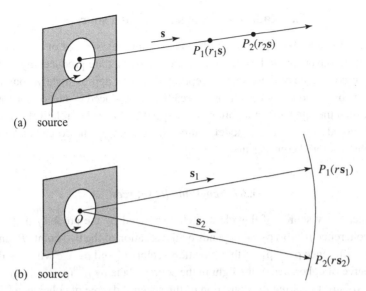

Fig. 5.7 Illustrating longitudinal spectral coherence (a) and transverse spectral coherence (b) in the far zone.

We note that $\mu^{(\infty)}$ depends on the distances r_1 and r_2 from the origin of the two points $P_1(r_1\mathbf{s})$ and $P_2(r_2\mathbf{s})$ in the far zone only through the phase factor $k(r_2 - r_1)$.

Two consequences of Eq. (22) are of special interest. When the points P_1 and P_2 in the far zone are located in the same direction, i.e. when $\mathbf{s}_2 \equiv \mathbf{s}_1$, the first factor on the right-hand side of Eq. (22) has the value unity and one then has

$$\mu^{(\infty)}(r_1\mathbf{s}, r_2\mathbf{s}, \omega) = e^{ik(r_2-r_1)}. \tag{23}$$

Consequently

$$\left|\mu^{(\infty)}(r_1\mathbf{s}, r_2\mathbf{s}, \omega)\right| = 1. \tag{24}$$

This formula implies that along any direction pointing from the source, the far field is spatially completely coherent at each frequency ω, a result which may be expressed by saying that *the far field has complete longitudinal spectral coherence at each frequency, irrespective of the state of coherence of the source* [Fig. 5.7(a)].

Let us next consider the situation in which the two points in the far zone are located at the same distance, r say, from an origin in the source region, in directions $\mathbf{s}_1$ and $\mathbf{s}_1$. The spectral degree of coherence $\mu^{(\infty)}(r\mathbf{s}_1, r\mathbf{s}_2, \omega)$ may then be said to represent *transverse coherence*. It follows at once from the general formula (21) that *the transverse degree of coherence of the light in the far zone is independent of the distance r and depends only on the directions* $\mathbf{s}_1$ *and* $\mathbf{s}_2$ [Fig. 5.7(b)].

5.3 Radiation from some model sources

To gain some insight about properties of fields radiated by sources of different states of coherence it is convenient and useful to consider fields that are produced by certain types of model sources. Some of these sources represent, at least approximately, sources that are frequently encountered in nature or that can readily be produced in a laboratory. Moreover, the fields which they generate may often be analyzed by the use of relatively simple mathematics. A broad class of such model sources is formed by the so-called *Schell-model sources*, which we will consider first.

5.3.1 Schell-model sources

A planar secondary source of this class is characterized by the property that its spectral degree of coherence $\mu^{(0)}(\rho_1, \rho_2, \omega)$ depends on the location of the two points P_1 and P_2 only through the difference $\rho_2 - \rho_1$, of their position vectors ρ_1 and ρ_2. We will then denote the spectral degree of coherence of the light in the source plane by $\mu^{(0)}(\rho_1' - \rho_1', \omega)$ rather than by $\mu^{(0)}(\rho_1, \rho_2, \omega)$. Recalling the definition of the spectral degree of coherence [Eq. (6b) of Section 4.2], it follows that the cross-spectral density function of a planar, secondary Schell-model source has the form

$$W^{(0)}(\rho_1', \rho_2', \omega) = \sqrt{S^{(0)}(\rho_1', \omega)}\sqrt{S^{(0)}(\rho_2', \omega)}\mu^{(0)}(\rho_2' - \rho_1', \omega). \tag{1}$$

An expression for the radiant intensity of the field generated by a source of this kind may readily be calculated on first taking the four-dimensional spatial Fourier transform of expression (1) and substituting it into the general formula [Eq. (20) of Section 5.2]. On changing the variables of integration in the Fourier transform by setting $\rho_1' + \rho_2' = 2\rho$ and $\rho_2' - \rho_1' = \rho'$ one obtains the following expression for the radiant intensity generated by such a source:

$$J_\omega(s) \equiv J_\omega(\theta) = \left(\frac{k}{2\pi}\right)^2 \cos^2\theta \int \mu^{(0)}(\rho', \omega)H^{(0)}(\rho', \omega)e^{-iks_\perp \cdot \rho'} \, d^2\rho', \tag{2}$$

where

$$H^{(0)}(\rho', \omega) = \int \sqrt{S^{(0)}(\rho + \rho'/2, \omega)}\sqrt{S^{(0)}(\rho - \rho'/2, \omega)} \, d^2\rho. \tag{3}$$

As an example let us suppose that the source is circular and of radius a and that both the spatial distribution of its spectral density across it and also its spectral degree of coherence are Gaussian, i.e. that

$$S^{(0)}(\rho, \omega) = \begin{cases} A^2 e^{-\rho^2/(2\sigma_s^2)} & \text{when } \rho \leq a \\ 0 & \text{when } \rho > a \end{cases} \tag{4}$$

and

$$\mu^{(0)}(\rho',\omega) = e^{-\rho'^2/(2\sigma_\mu^2)}, \tag{5}$$

where the quantities A, σ_s and σ_μ are taken to be independent of position but, in general, depend on the frequency. Usually $a \gg \sigma_s$. We will assume this to be the case, which simplifies subsequent calculations. Such sources are known as *Gaussian Schell-model sources* and the fields which they generate are called *Gaussian Schell-model fields*. They have been studied extensively in the analysis of various partially coherent fields and in connection with propagation of beams through the turbulent atmosphere where fields of this kind, which have beam-like form, are called *Gaussian Schell-model beams*. We will encounter such beams in Chapter 8.

On substituting from Eqs. (3)–(5) into Eq. (2) one finds after a long but straightforward calculation that

$$H^{(0)}(\rho',\omega) = 2\pi\, A^2\sigma_s^2 e^{-\rho'^2/(8\sigma_s^2)}, \tag{6}$$

and

$$J_\omega(\theta) = J_\omega(0)\cos^2\theta\, e^{-\frac{1}{2}(k\delta)^2\sin^2\theta}, \tag{7}$$

where we have now written $J_\omega(\theta)$ rather than $J_\omega(s)$, θ denoting, as before, the angle which the unit vector s makes with the positive z-axis; and we assumed not only that $a \gg \sigma_s$ but also that $a \gg \sigma_\mu$. In Eq. (7)

$$J_\omega(0) = (kA\sigma_s\delta)^2 \tag{8}$$

and

$$\frac{1}{\delta^2} = \frac{1}{(2\sigma_s)^2} + \frac{1}{\sigma_\mu^2}. \tag{8a}$$

We see from Eq. (5) that $\mu^{(0)}(\rho',\omega) \rightarrow 1$ as $\sigma_\mu \rightarrow \infty$. The source is then fully spatially coherent, but this limit has to be interpreted with caution. For our earlier assumption that $a \gg \sigma_\mu$ demands that the source diameters must also become infinite, in such a manner that the ratio a/σ_μ tends to a finite limit which is large relative to unity. Now according to Eq. (4) the spatial intensity distribution across the source has a Gaussian form. This is clearly the same situation as one encounters when the source is a laser operating in its lowest-order Hermite–Gaussian mode. In this case ($\sigma_\mu \rightarrow \infty$) Eq. (8a) implies that

$$\delta = 2\sigma_s. \tag{9}$$

Moreover, we now have $k\delta \equiv 2\pi(\delta/\lambda) = 4\pi\sigma_s/\lambda$ and, for a realistic laser, this parameter will be much greater than unity. The exponential term in Eq. (7) will then have a non-negligible value only when $\sin\theta \ll 1$. The expression (7) for the radiant intensity $J_\omega(\theta)$ generated by such a source may then be approximated by the expression

$$J_\omega(\theta) = J_\omega(0)e^{-\frac{1}{2}(k\delta)^2\theta^2}.$$ (10)

Evidently the angular spread of the radiation from such a source is confined to a domain of semi-angle $\theta \sim 1/(k\delta) = \lambda/(2\pi\delta) \sim \lambda/(4\pi\sigma_s)$, in view of Eq. (9). In this case the radiation is effectively confined to a very narrow angular domain, i.e. under these conditions the source generates a beam – namely a Gaussian beam, with a beam waist $w_0 = \delta = 2\sigma_s$. Thus we have shown that Gaussian Schell-model sources are generalizations of laser sources operating in their lowest-order Hermite–Gaussian mode.

Let us now consider the other extreme case, namely when $\sigma_\mu \to 0$. This obviously represents the *incoherent limit* (zero spectral correlation length). In this case we have from Eq. (8a) that $\delta \to 0$. In order for the factor $J_\omega(0)$ given by the formula (8) to remain finite and non-zero, with σ_s fixed, it is necessary that $A \to \infty$ in such a way that the product $A\delta$ remains finite. Expression (7) then reduces to

$$J_\omega(\theta) = J_\omega(0)\cos^2\theta.$$ (11)

Hence in the incoherent limit the radiant intensity falls off with θ as $\cos^2\theta$. Since, for a Lambertian source, $J_\omega(\theta) = J_\omega(0)\cos\theta$, this result implies that a spatially strictly incoherent planar, secondary source cannot be Lambertian, at least when the incoherent source is modeled as a limiting case of a Gaussian Schell-model source. One might, therefore, suspect that a Lambertian source is not completely spatially incoherent. We will later see (Section 5.5) that this is indeed the case.

5.3.2 Quasi-homogeneous sources

An important sub-class of Schell-model sources (again assumed to be planar, secondary sources) is constituted by so-called *quasi-homogeneous* sources. For such sources the spectral density $S^{(0)}(\rho, \omega)$ changes much more slowly with ρ than the spectral degree of coherence $\mu^{(0)}(\rho', \omega)$ changes with $\rho' = \rho_2' - \rho_1'$. One says that $S^{(0)}(\rho, \omega)$ is a "slow" function of ρ whereas $\mu^{(0)}(\rho', \omega)$ is a "fast" function of ρ'. Such behavior is illustrated in Fig. 5.8. Further, the linear dimensions of sources of this kind are assumed to be large relative to the wavelength $\lambda = 2\pi c/\omega$ of the light.

For a quasi-homogeneous source the expression (1) for the cross-spectral density simplifies. It may evidently be approximated by the formula

$$W^{(0)}(\rho_1', \rho_2', \omega) \approx S^{(0)}\left(\frac{\rho_1' + \rho_2'}{2}, \omega\right)\mu^{(0)}(\rho_2' - \rho_1', \omega).$$ (12)

The fact that $W^{(0)}$ now has such a factorized form leads to considerable simplification in further analysis.

To derive an expression for the radiant intensity and for the spectral degree of coherence of the far field generated by a planar, quasi-homogeneous source we must first determine the

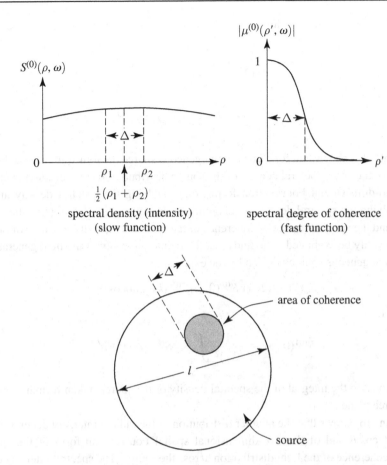

Fig. 5.8 Illustrating the concept of a quasi-homogeneous source. The effective spectral coherence area of the source is much smaller than the region of the source over which the spectral density (intensity) changes appreciably.

four-dimensional spatial Fourier transform of the expression (12). For this purpose we introduce the same new variables as we did before in the transition from Eq. (1) to Eq. (2), viz.,

$$\rho = \tfrac{1}{2}(\rho_1' + \rho_2'), \qquad \rho' = \rho_2' - \rho_1'. \tag{13a}$$

We have the inverse relations

$$\rho_1' = \rho - \tfrac{1}{2}\rho', \qquad \rho_2' = \rho + \tfrac{1}{2}\rho', \tag{13b}$$

and one readily finds that the four-dimensional Fourier transform $\tilde{W}^{(0)}$, defined by Eq. (14) of Section 5.2, of the cross-spectral density function (12) has the factorized form

$$\tilde{W}^{(0)}(\mathbf{f}_1, \mathbf{f}_2, \omega) = \tilde{S}^{(0)}(\mathbf{f}_1 + \mathbf{f}_2, \omega)\tilde{\mu}^{(0)}\left[\tfrac{1}{2}(\mathbf{f}_2 - \mathbf{f}_1), \omega\right], \tag{14}$$

where $\tilde{S}^{(0)}$ and $\tilde{\mu}^{(0)}$ are the two-dimensional Fourier transforms of $S^{(0)}$ and of $\mu^{(0)}$, respectively:

$$\tilde{S}^{(0)}(\mathbf{f}, \omega) = \frac{1}{(2\pi)^2} \int_{(z=0)} S^{(0)}(\boldsymbol{\rho}, \omega) e^{-i\mathbf{f} \cdot \boldsymbol{\rho}} d^2\rho, \tag{15a}$$

$$\tilde{\mu}^{(0)}(\mathbf{f}', \omega) = \frac{1}{(2\pi)^2} \int_{(z=0)} \mu^{(0)}(\boldsymbol{\rho}', \omega) e^{-i\mathbf{f}' \cdot \boldsymbol{\rho}'} d^2\rho'. \tag{15b}$$

The formula (14) shows that for a quasi-homogeneous source the four-dimensional Fourier transform of the cross-spectral density of the source distribution factorizes into the product of two two-dimensional Fourier transforms, one involving the spectral density and the other involving the spectral degree of coherence of the source. Using Eq. (14), the radiant intensity and the spectral degree of coherence of the far field [Eqs. (20) and (22) of Section 5.2] may readily be evaluated. One finds that the radiant intensity of the field generated by a quasi-homogeneous source is given by the expression

$$J_\omega(\theta) = (2\pi k)^2 \tilde{S}^{(0)}(0, \omega) \tilde{\mu}^{(0)} (k\mathbf{s}_\perp, \omega) \cos^2\theta, \tag{16}$$

where the factor

$$\tilde{S}^{(0)}(0, \omega) = \frac{1}{(2\pi)^2} \int_{(z=0)} S^{(0)}(\boldsymbol{\rho}', \omega) d^2\rho' \tag{17}$$

is proportional to the integral of the spectral density of the source, taken formally over the entire source plane.

Equation (16) shows that the angular distribution of the radiant intensity depends on the product of $\cos^2\theta$ and of the two-dimensional spatial Fourier transform of the spectral degree of coherence of the light distribution across the source. The spectral intensity distribution across the source, i.e. the spectral density $S^{(0)}(\rho, \omega)$, enters only through the factor $\tilde{S}^{(0)}(0, \omega)$ given by Eq. (17). Hence *the angular distribution of radiant intensity generated by a quasi-homogeneous source is independent of the shape of the source*; it is essentially determined by the spatial-coherence properties of the source, represented by the spectral degree of coherence $\mu^{(0)}$. This result explains a fact we noted earlier, in Section 5.1, namely that *the angular distribution of the intensity throughout the far zone generated by Lambertian sources (which, as we will learn in Section 5.5, are quasi-homogeneous) is independent of the shape of the source.*

The far-zone coherence properties of the radiation produced by a planar, secondary quasi-homogeneous source may also be readily determined. We see from Eq. (14) that

$$\tilde{W}^{(0)}(-k\mathbf{s}_{1\perp}, k\mathbf{s}_{2\perp}, \omega) = \tilde{S}^{(0)}[k(\mathbf{s}_{2\perp} - \mathbf{s}_{1\perp}), \omega] \tilde{\mu}^{(0)} \left[\tfrac{1}{2} k(\mathbf{s}_{1\perp} + \mathbf{s}_{2\perp}), \omega\right], \tag{18}$$

and there are two similar expressions for the two terms appearing in the denominator of Eq. (22) of Section 5.2 for the spectral degree of coherence. On substituting from Eq. (18) into that equation we find that

$$\mu^{(\infty)}(r_1\mathbf{s}_1, r_2\mathbf{s}_2, \omega) = \frac{\tilde{S}^{(0)}[k(\mathbf{s}_{2\perp} - \mathbf{s}_{1\perp}), \omega]}{\tilde{S}^{(0)}(0, \omega)} \tilde{G}^{(0)}(k\mathbf{s}_{1\perp}, k\mathbf{s}_{2\perp}, \omega) e^{ik(r_2 - r_1)}, \qquad (19)$$

where

$$\tilde{G}^{(0)}(k\mathbf{s}_{1\perp}, k\mathbf{s}_{2\perp}, \omega) = \frac{\tilde{\mu}^{(0)}\left[\frac{1}{2} k(\mathbf{s}_{1\perp} + \mathbf{s}_{2\perp}), \omega\right]}{\sqrt{\tilde{\mu}^{(0)}(k\mathbf{s}_{1\perp}, \omega)}\sqrt{\tilde{\mu}^{(0)}(k\mathbf{s}_{2\perp}, \omega)}}. \qquad (20)$$

Now, for a quasi-homogenous source, the spectral density $S^{(0)}(\rho, \omega)$ is a "slow" function of ρ whereas $\mu^{(0)}(\rho', \omega)$ is a "fast" function of ρ'. According to a well-known reciprocity relation involving Fourier-transform pairs, $\tilde{S}^{(0)}(\mathbf{f}, \omega)$ will, therefore, be a "fast" function of $\mathbf{f}$ and $\tilde{\mu}^{(0)}(\mathbf{f}', \omega)$ will be a "slow" function of $\mathbf{f}'$. Using these facts, the expression on the right of Eq. (20) may be approximated by unity and the expression (19) for the spectral degree of coherence of the far field reduces to

$$\mu^{(\infty)}(r_1\mathbf{s}_1, r_2\mathbf{s}_2, \omega) = \frac{\tilde{S}^{(0)}[k(\mathbf{s}_{2\perp} - \mathbf{s}_{1\perp}), \omega]}{\tilde{S}^{(0)}(0, \omega)} e^{ik(r_2 - r_1)}. \qquad (21)$$

This formula shows that, apart from a simple geometrical phase factor, the spectral degree of coherence $\mu^{(\infty)}(r_1\mathbf{s}_1, r_2\mathbf{s}_2, \omega)$ of the far field generated by a planar, secondary quasi-homogeneous source is equal to the normalized spatial Fourier transform of the spectral density across the source.

Except for the difference in notation, the formula (21) is of exactly the same form as the far-zone form of the van Cittert–Zernike theorem [Eq. (20) of Section 3.2] for the equal-time degree of coherence $\gamma^{(\infty)}(r_1\mathbf{s}_1, r_2\mathbf{s}_2, 0) \equiv j^{(\infty)}(r_1\mathbf{s}_1, r_2\mathbf{s}_2)$, which pertains to quasi-monochromatic radiation from a spatially incoherent source. Clearly the formula (21) may be regarded as its generalization, in the space–frequency domain, to the far field generated by quasi-homogeneous sources. It should be noted that such sources may have large coherence areas.

We will now examine more closely the two main formulas which we derived in this section, namely Eqs. (16) and (21), taking for simplicity $r_2 = r_1$ in Eq. (21). We may express these two formulas in the form

$$J_\omega(\theta) = (2\pi k)^2 C\tilde{\mu}^{(0)}(k\mathbf{s}_\perp, \omega)\cos^2\theta, \qquad (16a)$$

$$\mu^{(\infty)}(r_1\mathbf{s}_1, r_2\mathbf{s}_2, \omega) = \frac{1}{C} \tilde{S}^{(0)}[k(\mathbf{s}_{2\perp} - \mathbf{s}_{1\perp}), \omega], \qquad (21a)$$

where the factor $C \equiv \tilde{S}^{(0)}(0, \omega)$ is given by Eq. (17). This pair of formulas brings into evidence two interesting *reciprocity theorems* relating to radiation from planar quasi-homogeneous sources. The first shows that *the angular distribution of the radiant intensity primarily depends on the two-dimensional spatial Fourier transform of the spectral degree of coherence of light across the source and, as already noted, is therefore independent of the shape of the source.* The second shows that *the spectral degree of coherence of the far field is proportional to the two-dimensional spatial Fourier transform of the spectral intensity distribution (the spectral density) across the source.* These two results are indicated schematically in the following diagram, in which the arrowed lines indicate Fourier-transform pairs:

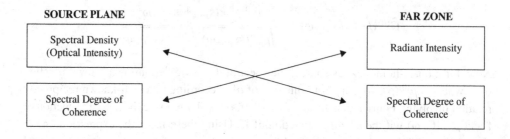

We will illustrate these results by a simple example. Consider radiation from a secondary, quasi-homogeneous uniform circular source of radius $a \gg \lambda$ whose spatial distribution of the spectral intensity and also spectral degree of coherence are Gaussian, viz.,

$$S^{(0)}(\boldsymbol{\rho}, \omega) = \begin{cases} A^2 e^{-\rho^2/(2\sigma_s^2)} & \text{when } \rho \leq a, \\ 0 & \text{when } \rho > a, \end{cases} \tag{22}$$

and

$$\mu^{(0)}(\boldsymbol{\rho}', \omega) = e^{-\rho'^2/(2\sigma_\mu^2)}, \tag{23}$$

where the parameters σ_s and σ_μ depend on frequency, in general. The calculations simplify if we assume that the linear dimensions of the source are large relative to the r.m.s. width σ_s of the spectral density function $S^{(0)}$, i.e. if $a \gg \sigma_s$. Because the source is assumed to be quasi-homogeneous, we also have $a \gg \sigma_\mu$. Under these circumstances one readily obtains from the two reciprocity relations (16a) and (21a) the following expressions for the radiant intensity and spectral degree of coherence of the far field generated by such a source:

$$J_\omega(\theta) = (kA\sigma_\mu\sigma_s)^2 \cos^2\theta \, e^{-\frac{1}{2}[(k\sigma_\mu)^2 \sin^2\theta]}, \tag{24}$$

$$\mu^{(\infty)}(r_1\mathbf{s}_1, r_2\mathbf{s}_2, \omega) = e^{-\frac{1}{2}(k\sigma_s)^2 u_{12}^2} e^{ik(r_2-r_1)}, \tag{25}$$

where

$$u_{12} = |\mathbf{s}_{2\perp} - \mathbf{s}_{1\perp}|. \tag{26}$$

Figure 5.9 shows the polar diagram of the normalized radiant intensity $J_\omega(\theta)/J_\omega(0)$, calculated from Eq. (24), for radiation from sources with various values of the r.m.s. width,

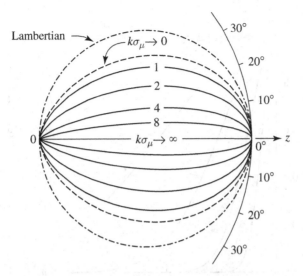

Fig. 5.9 Polar diagrams, calculated from Eq. (24) of Section 5.3, of the normalized radiant intensity from a Gaussian-correlated quasi-homogeneous source, for various values of the r.m.s. width σ_μ of the spectral degree of coherence μ. The length of the vector pointing from the origin to a typical point on a curve labeled by a particular value of the parameter $k\sigma_\mu$ represents the normalized radiant intensity in the direction of that vector. [Adapted from E. Wolf and W. H. Carter, *Opt. Commun.* **13** (1975), 205–206.]

σ_μ, of the spectral degree of coherence μ. We see that for small values of $k\sigma_\mu$ the radiation is spread over a wide solid angle but that as $k\sigma_\mu$ increases, i.e. as the source becomes more spatially coherent, the radiation becomes more directional, eventually forming a beam. For comparison, the corresponding polar diagram for radiation from a Lambertian source is included. Figure 5.10 illustrates the behavior of the spectral degree of coherence of the light in the far zone, calculated from Eq. (25). It shows that with increasing effective source size ($k\sigma_s$ increasing) the angular region over which the absolute value of the degree of coherence is appreciable becomes narrower.

5.4 Sources of different states of spatial coherence which generate identical distributions of the radiant intensity

It is frequently asserted that, in order for a source to generate a highly directional field, i.e. a narrow beam of radiation, it must be spatially highly coherent, such as, for example, a well-stabilized laser is. This, however, is not true. We will see that under appropriate conditions partially coherent sources of different states of coherence may produce beams which are just as directional as laser beams. More specifically we will show that sources of different states of coherence may produce fields with the same far-zone intensity distribution, i.e. with the same distribution of the radiant intensity.

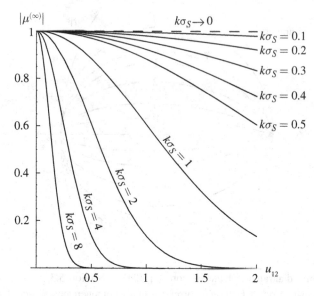

Fig. 5.10 The behavior of the absolute value of the spectral degree of coherence, given by Eq. (25) of Section 5.3, of the far field generated by a quasi-homogeneous, Gaussian, planar, secondary source. The variable u_{12} is defined by Eq. (26) of Section 5.4. [Adapted from W. H. Carter and E. Wolf, *J. Opt. Soc. Amer.* **67** (1977), 785–796.]

It follows from Eqs. (14) and (20) of Section 5.2 that the radiant intensity of a field generated by a planar, secondary, statistically stationary source may be expressed in the form

$$J_\omega(\mathbf{s}) = \left(\frac{k}{2\pi}\right)^2 \cos^2\theta \iint W^{(0)}(\boldsymbol{\rho}_1', \boldsymbol{\rho}_2', \omega) e^{-i k \mathbf{s}_\perp \cdot (\boldsymbol{\rho}_2' - \boldsymbol{\rho}_1')} \, d^2\rho_1' d^2\rho_2'. \tag{1}$$

Let us express the cross-spectral density $W^{(0)}$ of the field in the source plane in terms of the spectral density and the spectral degree of coherence of the light in that plane [Eq. (6b) of Section 4.2]. The formula (1) then becomes

$$J_\omega(\mathbf{s}) = \left(\frac{k}{2\pi}\right)^2 \cos^2\theta \iint \sqrt{S^{(0)}(\boldsymbol{\rho}_1', \omega)} \sqrt{S^{(0)}(\boldsymbol{\rho}_2', \omega)} \mu^{(0)}(\boldsymbol{\rho}_1', \boldsymbol{\rho}_2', \omega) e^{-i k \mathbf{s}_\perp \cdot (\boldsymbol{\rho}_2' - \boldsymbol{\rho}_1')} \, d^2\rho_1' d^2\rho_2'.$$

$$\tag{2}$$

Equation (2) shows that both the spectral intensity distribution $S^{(0)}(\boldsymbol{\rho}, \omega)$ and the spectral degree of coherence $\mu^{(0)}(\boldsymbol{\rho}_1', \boldsymbol{\rho}_2', \omega)$ of the light in the source plane contribute to the radiant intensity. The possibility exists that two different sources, one with spectral distribution $S_1^{(0)}(\boldsymbol{\rho}', \omega)$ and spectral degree of coherence $\mu_1^{(0)}(\boldsymbol{\rho}_1', \boldsymbol{\rho}_2', \omega)$ say, and the other with distributions $S_2^{(0)}(\boldsymbol{\rho}', \omega)$ and $\mu_2^{(0)}(\boldsymbol{\rho}_1', \boldsymbol{\rho}_2', \omega)$, may generate the same radiant intensity $J_\omega(\mathbf{s})$. To put it differently, it seems plausible that there might be a "trade-off" between the spectral density and the spectral degree of coherence in such a way that the two sources would give the same values of the integrals in Eq. (2) and, consequently, generate the same radiant intensity. It turns out that this is indeed possible. A simple example is provided by the Gaussian Schell-model sources which we discussed in Section 5.3. It follows at once from Eqs. (6)–(8) of that section that two sources of this kind, for which the r.m.s. widths σ_s of the spectral density and the r.m.s. width σ_μ of the spectral degree of coherence are such that the quantity δ, defined by Eq. (8a) of Section 5.3, viz.,

$$\frac{1}{\delta^2} = \frac{1}{(2\sigma_s)^2} + \frac{1}{\sigma_\mu^2}, \tag{3}$$

has the same value, will generate fields with the same relative angular distribution of the radiant intensity. Moreover, as is evident from Eqs. (7) and (8) of Section 5.3, if the factors A [see Eq. (4) of Section 5.3] are such that the expression

$$J_\omega(0) = (kA\sigma_s\delta)^2 \tag{4}$$

also has the same values, then not only the relative but also the actual values of the radiant intensities will be the same.

The fact that for two such "equivalent" sources the sum of the two terms on the right-hand side of Eq. (3) has to be the same indicates the trade-off between the contributions of the spectral density and of the spectral degree of coherence. Some computed curves showing such a trade-off are shown in Fig. 5.11. Experimental verifications of these theoretical predictions are illustrated in Fig. 5.13, obtained by use of the system shown in Fig. 5.12.

In Fig. 5.14, the changes of beam radii for propagation from sources with the same initial r.m.s. radius $\sigma_s = 0.1$ cm but with different spectral degrees, of coherence (a) and from sources with the same r.m.s. spectral degree of coherence but with different r.m.s. widths σ_s of the intensity distribution (b) are shown. Figure 5.14(c) illustrates the equivalence theorem. There is a trade-off between spatial coherence of the source and the spatial intensity distribution across the source, in accordance with Eq. (3), resulting in beams which have the same angular spread in the far zone.

5.5 Coherence properties of Lambertian sources

We will now return to one of the reciprocity relations which we derived in Section 5.3 and make use of it to elucidate the spatial coherence properties of Lambertian sources.

According to the first reciprocity relation, given by Eq. (16) of Section 5.3, the radiant intensity of a field generated by a planar, secondary, quasi-homogeneous source is given by the expression

$$J_\omega(\mathbf{s}) = (2\pi k)^2 \, \tilde{S}^{(0)}(0,\omega)\tilde{\mu}^{(0)}(k\mathbf{s}_\perp,\omega)\cos^2\theta, \tag{1}$$

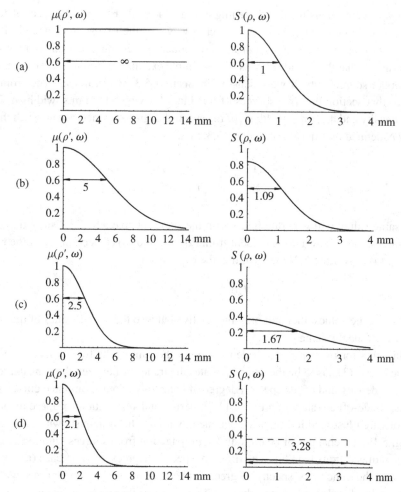

Fig. 5.11 Illustrating the behavior of the spectral degree of coherence and the spectral intensity distribution across three partially coherent sources [(b), (c), (d)] which produces fields whose far-zone intensity distributions are the same as that generated by a coherent laser source [(a)]. The parameters characterizing the four sources are (a) $\sigma_\mu = \infty$, $\sigma_S = 1$ mm, $A = 1$ (arbitrary units), (b) $\sigma_\mu = 5$ mm, $\sigma_S = 1.09$ mm, $A = 0.84$, (c) $\sigma_\mu = 2.5$ mm, $\sigma_S = 1.67$ mm, $A = 0.36$ and (d) $\sigma_\mu = 2.1$ mm, $\sigma_S = 3.28$ mm, $A = 0.09$. The normalized radiant intensity generated by all these sources is $J_\omega(\theta)/J_\omega(0) = \cos^2\theta \exp[-2(k\delta_L)^2 \sin^2\theta]$, ($\sigma_S = 1$ mm). [Adapted from E. Wolf and E. Collett, *Opt. Commun.* **25** (1978), 293–296.]

where $\tilde{S}^0(0, \omega)$ is defined by Eq. (17) of that section. For a Lambertian source,

$$J_\omega(\mathbf{s}) = J_\omega^{(0)} \cos \theta, \qquad (2)$$

where $J_\omega^{(0)}$ denotes the radiant intensity generated by the source in the direction $\theta = 0$.

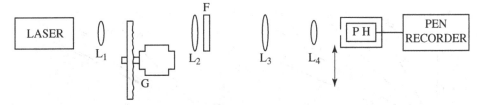

Fig. 5.12 A system that was used to test that sources with different spatial coherence properties can generate identical angular distributions of the radiant intensity. L_1, L_2, L_3 and L_4 are lenses, F is an amplitude filter, G is a rotating ground glass plate, PH is a photo-detector. [Adapted from P. DeSantis, F. Gori, G. Guattari and C. Palma, *Opt. Commun.* **29** (1979), 256–260.]

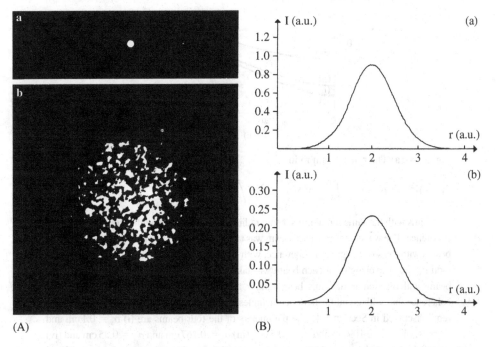

Fig. 5.13 (A) Intensity distributions across a coherent laser source (a) and across an "equivalent" partially coherent source (b). (B) The measured angular distribution of intensity, I (arbitrary but same units) in the far zone of fields generated by the two sources shown on the left. [Reproduced from P. DeSantis, F. Gori, G. Guattari and C. Palma, *Opt. Commun.* **29** (1979), 256–260.]

Suppose, as is usually the case, that the spectrum is the same at all source points, i.e. that $S^{(0)}(\rho', \omega) \equiv S^{(0)}(\omega)$ say. The formula (17) of Section (5.3) then gives at once

$$\tilde{S}^{(0)}(0, \omega) = \frac{A}{(2\pi)^2} S^{(0)}(\omega),$$ (3)

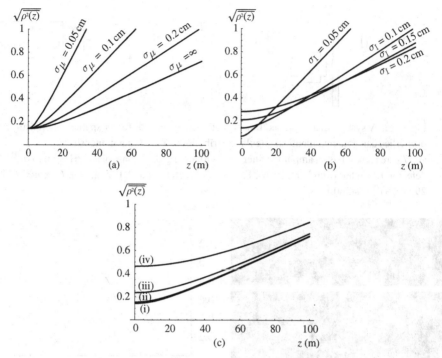

Fig. 5.14 (a) The r.m.s. beam radii

$$\left[\overline{\rho^2(z)}\right]^{\frac{1}{2}}, \quad \overline{\rho^2(z)} = \iint_{-\infty}^{\infty} \rho^2 S(\rho, z) \mathrm{d}^2\rho \bigg/ \iint_{-\infty}^{\infty} S(\rho, z) \mathrm{d}^2\rho,$$

for beams with the same initial r.m.s. beam radii ($\sigma_s = 0.1$ cm), but with different degrees of coherence. The wavelength for each beam was taken as 6328 Å. (b) The r.m.s. beam radii for beams with the same degree of coherence, with $\sigma_\mu = 0.2$ cm, but with different initial r.m.s. radii σ_s. The wavelength for each beam was taken as 6328 Å. (c) The r.m.s. beam radii for beams with different initial r.m.s. beam radii σ_s and with different widths σ_μ of the degree of coherence, but with equal far-field beam angles, θ_B, as predicted by the "equivalence theorem" discussed in Section 5.4. The parameters of the four beams are (i) $\sigma_s = 0.1$ cm and $\sigma_\mu = \infty$, (ii) $\sigma_s = 0.109$ cm and $\sigma_\mu = 0.5$ cm, (iii) $\sigma_s = 0.167$ cm and $\sigma_\mu = 0.25$ cm and (iv) $\sigma_s = 0.328$ cm and $\sigma_\mu = 0.21$ cm. The wavelength of each beam was taken as 6328 Å. [Adapted from J. T. Foley and M. S. Zubairy, *Opt. Commun.* **26** (1978), 297–300.]

where A denotes the area of the source. If we next substitute from Eq. (3) into Eq. (1), use Eq. (2) and solve for $\tilde{\mu}^{(0)}$, we find at once that

$$\tilde{\mu}^{(0)}(k s_\perp, \omega) = \frac{J_\omega^{(0)}}{A k^2 S^{(0)}(\omega)\sqrt{1 - s_\perp^2}}, \qquad (4)$$

where we have used the relation $\cos\theta = \sqrt{1 - s_\perp^2}$.

Next we take the two-dimensional spatial Fourier transform of Eq. (4) and neglect the contribution from the domain $|\mathbf{s}_\perp| > 1$, which is associated with evanescent waves whose contributions are negligible unless $\mu^{(0)}$ varies rapidly on a scale of the order of a wavelength. We then obtain for $\mu^{(0)}$ the expression

$$\mu^{(0)}(\boldsymbol{\rho}', \omega) \approx \frac{J_\omega^{(0)}}{Ak^2 S^{(0)}(\omega)} \int_{\mathbf{s}_\perp^2 \leq 1} \frac{1}{\sqrt{1 - \mathbf{s}_\perp^2}} e^{ik\mathbf{s}_\perp \cdot \boldsymbol{\rho}'} d^2(k\mathbf{s}_\perp). \tag{5}$$

Now since $\mu^{(0)}(0, \omega) = 1$ one has from Eq. (5)

$$1 \approx \frac{J_\omega^{(0)}}{Ak^2 S^{(0)}(\omega)} \int_{\mathbf{s}_\perp^2 \leq 1} \frac{1}{\sqrt{1 - \mathbf{s}_\perp^2}} d^2(k\mathbf{s}_\perp). \tag{6}$$

From Eqs. (5) and (6) it follows at once that the spectral degree of coherence of the source may be expressed in the form

$$\mu^{(0)}(\boldsymbol{\rho}', \omega) = \frac{F(\boldsymbol{\rho}', \omega)}{F(0, \omega)}, \tag{7}$$

where

$$F(\boldsymbol{\rho}', \omega) = \int_{\mathbf{s}_\perp^2 \leq 1} \frac{1}{\sqrt{1 - \mathbf{s}_\perp^2}} e^{ik\mathbf{s}_\perp \cdot \boldsymbol{\rho}'} d^2(k\mathbf{s}_\perp). \tag{8}$$

The integral on the right may be evaluated in closed form and one finds that (M&W, p. 248)

$$\mu^{(0)}(\boldsymbol{\rho}', \omega) = \frac{\sin(k\rho')}{k\rho'}. \tag{9}$$

This formula implies that *all quasi-homogeneous planar secondary Lambertian sources for which the spectrum is the same at each source point have the same spectral degree of coherence, given by Eq. (9)*. It is plotted in Fig. 5.15 as a function of the normalized distance $k\rho' \equiv k|\boldsymbol{\rho}'_2 - \boldsymbol{\rho}'_1|$ between two arbitrary source points. We see that the correlation distance $\Delta\rho'_c$ of the light in the source plane is given by the order of magnitude expression $(k\Delta\rho'_c) \approx \pi/2$ or, since $k = 2\pi/\lambda$,

$$\Delta\rho'_c \approx \lambda/4. \tag{10}$$

Thus we have shown that *Lambertian sources are not spatially completely incoherent but are correlated over distances of the order of the wavelength.*

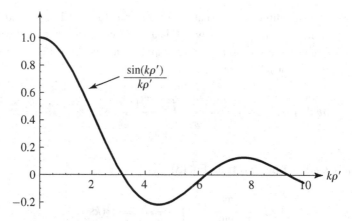

Fig. 5.15 The spectral degree of coherence across a planar, quasi-homogeneous secondary Lambertian source.

5.6 Spectral changes on propagation. The scaling law

It is generally taken for granted that the spectrum of light remains unchanged on propagation in free space. In 1986 it was discovered that, in general, the spectrum may change as light propagates,[1] such changes being caused by the spatial coherence properties of the light source. In this section we will discuss this phenomenon and we will also derive a condition which the spectral degree of coherence of the source has to satisfy in order that the normalized far-zone spectrum is the same as the spectrum across the source. We will consider only spectral changes in the far zone, produced by quasi-homogeneous sources.

The far-zone spectrum of light generated by a planar, secondary, quasi-homogeneous source is, according to Eq. (19) of Section 5.2 and Eq. (16) of Section 5.3, given by the expression

$$S^{(\infty)}(r\mathbf{s}, \omega) = \left(\frac{k}{r}\right)^2 \left(\int_{z=0} S^{(0)}(\boldsymbol{\rho}', \omega) \mathrm{d}^2 \rho' \right) \tilde{\mu}^{(0)}(k\mathbf{s}_\perp, \omega) \cos^2\theta. \tag{1}$$

If the spectrum of the source is the same at each source point, i.e. if $S^{(0)}(\boldsymbol{\rho}', \omega) \equiv S^{(0)}(\omega)$, Eq. (1) becomes

$$S^{(\infty)}(r\mathbf{s}, \omega) = \left(\frac{k}{r}\right)^2 A S^{(0)}(\omega) \tilde{\mu}^{(0)}(k\mathbf{s}_\perp, \omega) \cos^2\theta, \tag{2}$$

where, as before, A denotes the area of the source.

[1] E. Wolf, *Phys. Rev. Lett.* **56** (1986), 1370–1372. For a review of many of the publications on this subject see E. Wolf and D. F. V. James, *Rev. Prog. Phys.* **59** (1996), 771–818.

The normalized far-zone spectrum is, therefore,

$$s^{(\infty)}(r\mathbf{s}, \omega) \equiv \frac{S^{(\infty)}(r\mathbf{s}, \omega)}{\int_0^\infty S^{(\infty)}(r\mathbf{s}, \omega)d\omega}$$

$$= \frac{k^2 S^{(0)}(\omega)\tilde{\mu}^{(0)}(k\mathbf{s}_\perp, \omega)}{\int_0^\infty k^2 S^{(0)}(\omega)\tilde{\mu}^{(0)}(k\mathbf{s}_\perp, \omega)d\omega} \tag{3}$$

and the normalized source spectrum is

$$s^{(0)}(r\mathbf{s}, \omega) = \frac{S^{(0)}(\omega)}{\int_0^\infty S^{(0)}(\omega)d\omega}. \tag{4}$$

It is clear from a comparison of Eqs. (3) and (4) that, in general, the two normalized spectra will differ from each other, the difference being due to the spatial-coherence properties of the source, which are characterized by the spectral degree of coherence, $\mu^{(0)}$.

Examples of such "correlation-induced" spectral changes are given in Fig. 5.16, for radiation from a source with a Gaussian spectral profile. Figure 5.16 shows the differences in the far-zone spectrum in different directions of observation. We see that with increasing angle of observation from the normal to the source plane the spectral line undergoes a redshift, i.e. a shift towards longer wavelengths (lower frequencies). Other types of changes may be generated with different spatial-coherence properties of the source as is illustrated, for example, by Fig. 5.17.

It is clear from Eq. (3) that the normalized far-zone spectrum will be independent of the direction $\mathbf{s}$ of observation if $\tilde{\mu}(k\mathbf{s}_\perp, \omega)$ factorizes in the form

$$\tilde{\mu}^{(0)}(k\mathbf{s}_\perp, \omega) = F(\omega)\tilde{H}(\mathbf{s}_\perp). \tag{5}$$

The expression (3) then reduces to

$$s^{(\infty)}(r\mathbf{s}, \omega) = \frac{k^2 S^{(0)}(\omega)F(\omega)}{\int_0^\infty k^2 S^{(0)}(\omega)F(\omega)d\omega}. \tag{6}$$

The factorization condition (5) has some interesting consequences, if we assume that it holds not only for the domain $|s_\perp| \leq 1$ (which is a consequence of the fact that $\mathbf{s}$ is a unit vector) but also for all two-dimensional vectors $\mathbf{s}$. Evidently this will approximately be so, if at each frequency ω which is present in the source spectrum the spectral degree of coherence $\mu^{(0)}(\rho', \omega)$ is effectively bandlimited to a circle of radius k about the origin; or, in more physical terms, if $\mu^{(0)}(\rho', \omega)$ does not vary appreciably over distances of the order of the wavelength $\lambda = 2\pi c/\omega$. Assuming this to be the case, we obtain, on taking the Fourier transform of Eq. (5),

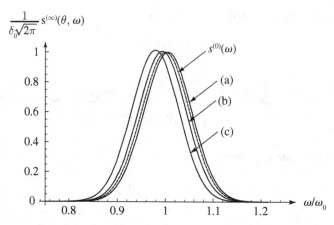

Fig. 5.16 Illustrating the effect of spatial coherence of a planar, secondary source on the normalized far-zone spectrum of the emitted light. The normalized source spectrum $s^{(0)}(\omega)$ is a line with a Gaussian profile, with $\sigma_s/\omega_0 = 1/20$ (ω_0 is the central frequency; σ_s is the r.m.s. width). The spectral degree of coherence was assumed to have a Gaussian spatial profile, with r.m.s. width $\sigma_\mu = 10\lambda_0$, λ_0 being the corresponding wavelength. The normalized far-zone spectrum is plotted (a) on axis, (b) at $\theta = 2°$ and (c) at $\theta = 30°$. [Adapted from Z. Dacic and E. Wolf, *J. Opt. Soc. Amer.* **A5** (1988), 1118–1126.]

$$\mu^{(0)}(\rho', \omega) = k^2 F(\omega) H(k\rho'), \tag{7}$$

where H is, of course, the Fourier transform of $\tilde{H}$. Since $\mu^{(0)}(\rho', \omega)$ is a correlation coefficient,

$$\mu^{(0)}(0, \omega) = 1 \qquad \text{for all } \omega. \tag{8}$$

Hence Eq. (7) implies that

$$k^2 F(\omega) = \frac{1}{H(0)}. \tag{9}$$

Because the left-hand side of Eq. (9) depends on the frequency but the right-hand side is independent of it, it follows that

$$F(\omega) = \frac{\alpha}{k^2}, \tag{10}$$

where α is a constant.

Two important conclusions follow at once from these results. If we substitute from Eq. (10) into Eq. (6), we obtain the following expression for the normalized far-field spectrum:

$$s^{(0)}(r\mathbf{s}, \omega) = \frac{S^{(0)}(\omega)}{\displaystyle\int_0^\infty S^{(0)}(\omega)d\omega}. \tag{11}$$

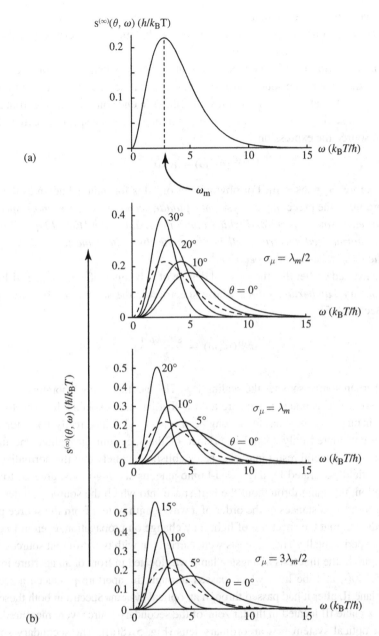

Fig. 5.17 The far-zone spectra in different directions of observation, produced by a planar, quasi-homogeneous source, with Gaussian spatial profile, whose spectrum is given
by Planck's law, (a) for the case when the spectral degree of coherence of the source is
a Gaussian function. (b) The frequency ω_m is the frequency at which the Planck spectrum, indicated by dotted line, has its maximum. [Adapted from E. Wolf, *Appl. Phys.*
B60 (1995), 303–308.]

105

This formula shows not only that the normalized spectrum of the light is now the same throughout the far zone, i.e. it is independent of the direction of observation, but also that it is equal to the normalized source spectrum.

When the condition (10) is satisfied and, consequently, when the normalized far-zone spectrum is the same as the source spectrum, the spectral degree of coherence necessarily has a certain functional form. This conclusion follows at once on substituting from Eq. (10) into Eq. (7) and setting $\alpha H(k\rho') = h(k\rho')$. We then obtain for the spectral degree of coherence of the source the expression

$$\mu^{(0)}(\rho', \omega) = h(k\rho'), \tag{12}$$

where, as before, $\rho' = \rho_2 - \rho_1$. For obvious reasons this formula is known as the *scaling law*. According to the preceding analysis, *any planar, secondary source whose spectrum is the same at each source point and which obeys the scaling law [Eq. (12)] will generate light whose normalized spectrum will be the same throughout the far zone and will be equal to the normalized source spectrum.*[1]

We have learned earlier that the spectral degree of coherence of a planar, quasi-homogeneous secondary *Lambertian source* whose spectrum is the same at each source point is [Eq. (9), Section 5.5]

$$\mu^{(0)}(\rho', \omega) = \frac{\sin(k\rho')}{k\rho'}, \tag{13}$$

i.e. a Lambertian source satisfies the scaling law. The fact that many laboratory sources and also many sources encountered in nature are Lambertian may explain why "spectral invariance" of light on propagation has for so long been generally, but incorrectly, taken for granted.

We have considered only spatial invariance of the spectrum throughout the far zone. However, a more general result may also be established, namely that the normalized spectrum of the field generated by any quasi-homogeneous *scaling-law* source is, to a good approximation, the same throughout the half-space into which the source radiates, except perhaps at points at distances of the order of a wavelength or less from the source plane.[2]

The prediction that the spectrum of light may change on propagation, even in free space, was tested experimentally. The first tests were carried out with two different sources, both of thermal origin. In the first tests a tungsten lamp was located in front of an aperture in plane I [see Fig. 5.18(a)] and the light which passed through the aperture produced a secondary source in plane II, after it had passed through an optical system. Spectra in both these planes and also in a plane III located in the far zone of the secondary source were measured.

The first optical system was an ordinary lens [Fig. 5.18(a)]. The secondary source in plane II can be shown to have a spectral degree of coherence which obeys the scaling law. The second system [Fig. 5.18(b)] was a Fourier achromat consisting of a combination of

[1] We established the condition (12) as a sufficiency condition for "spectral invariance." The condition is also a necessary condition [E. Wolf, *J. Mod. Opt.* **39** (1992), 9–20, Theorem II, p. 19].

[2] H. Roychowdhury and E. Wolf, *Opt. Commun.* **215** (2003), 199–203.

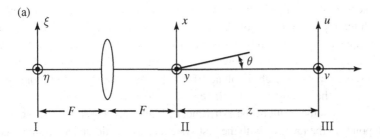

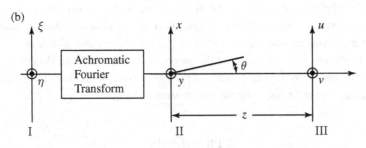

Fig. 5.18 Two systems used to illustrate the validity of the scaling law for a planar, secondary, quasi-homogeneous sources. [Adapted from G. M. Morris and D. Faklis, *Opt. Commun.* **62** (1987), 5–11.]

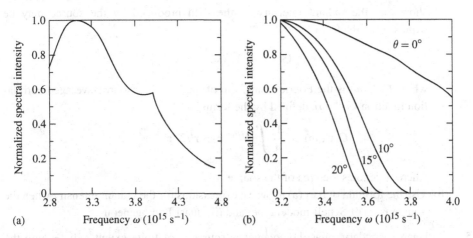

Fig. 5.19 (a) Observed spectra when the scaling law is satisfied. (b) Observed spectra when the scaling law is not satisfied. [Adapted from G. M. Morris and D. Faklis, *Opt. Commun.* **62** (1987), 5–11.]

optical elements designed for the use in white-light processing. One can show that with this system the light in plane II does not obey the scaling law; in fact it is effectively independent of the frequency over the whole frequency range for which the lens has been achromatized. According to the theory which we outlined, one must expect that in the first case the normalized spectrum of the light in plane III will be the same in all directions of observation θ, whereas in the second case it will depend on θ.

Figure 5.19 shows the result of the experiments. Figure 5.19(a) shows the measured spectrum at all points in the far zone with the first system (a conventional lens). Since, as we have already noted, the scaling law is satisfied in this case, the theory predicts that the normalized spectrum of the light is the same for all θ directions and is the same as the normalized spectrum of the secondary source. This is indeed what the experiments demonstrated. In Fig. 5.19(b) the measured far-zone spectra obtained in the experiment which used the Fourier achromat are shown. Since in this case the secondary source in plane II does not obey the scaling law, the normalized spectra observed in different directions should depend on θ. This was experimentally confirmed, with the results shown in Fig. 5.19(b).

PROBLEMS

5.1 A planar secondary source occupies a finite domain in the plane $z = 0$ and radiates into the half-space $z > 0$.

(a) If the source fluctuations are characterized by an ensemble which is stationary, show that the radiant intensity of the field produced by the source may be expressed in the form

$$J_\omega(\mathbf{s}) = k^2 A \tilde{C}(k\mathbf{s}_\perp, \omega)\cos^2\theta,$$

where $\tilde{C}(K, \omega)$ is the Fourier transform of the so-called source-averaged correlation function $C(\mathbf{r}', \omega)$, defined by the formula

$$C(\mathbf{r}', \omega) = \frac{1}{A} \int W^{(0)}(\mathbf{r} - \mathbf{r}'/2, \mathbf{r} + \mathbf{r}'/2)\mathrm{d}^2r,$$

where A denotes the area of the source.

(b) Use the formula of part (a) to derive expressions for the radiant intensity when the source is (i) of Schell-model type; and (ii) quasi-homogeneous.

5.2 A planar, secondary, quasi-homogeneous source σ of finite extent radiates into the half-space $z > 0$. The total radiated flux at frequency ω is given by the expression

$$F_\omega = \int_\sigma J_\omega(\mathbf{s})\mathrm{d}\Omega,$$

where $J_\omega(\mathbf{s})$ is the radiant intensity generated by the source and the integration extends over the solid angle of 2π subtended at the source by the hemisphere at infinity in the half-space $z > 0$. Show that

$$F_\omega \le \int_\sigma S^{(0)}(\mathbf{r}, \omega)d^2r,$$

where $S^{(0)}(\mathbf{r}, \omega)$ is the distribution of the spectral density across the source.

5.3 Obtain conditions under which two planar, secondary, Gaussian Schell-model sources will generate fields that have the same spectral degree of coherence throughout the far zone.

Derive also expressions for the corresponding radiant intensities produced by two such sources.

5.4 A three-dimensional spatially incoherent primary source occupies a finite domain D. Its fluctuations are statistically stationary and the spectrum of the source is the same at every point.
   (a) Show that the normalized spectrum of the field is the same at every point outside the source domain D.
   (b) Derive an expression for the spectral degree of coherence in the far zone.

5.5 (a) Find a sufficiency condition under which two three-dimensional statistically stationary, primary, quasi-homogeneous source distributions $Q(\mathbf{r}, \omega)$ generate fields which have the same far-zone spectra.
   (b) Give an example of two such sources.
   (c) Consider also the special case of (a) when the sources obey the scaling law, i.e. when the spectral degree of coherence of each source has the functional form

$$\mu_Q(\mathbf{r}', \omega) = h(kr'), \quad (k = \omega/c).$$

5.6 Consider a statistically stationary three-dimensional source occupying a finite volume.
   (a) Derive an expression for the radiant intensity in terms of the coherent modes of the source.
   (b) If the integral of the radiant intensity over all directions (i.e. directions that fill the complete 4π solid angle) has zero value, the source may be said to be non-radiating. Show that, if the source is non-radiating, then all of its source modes are also non-radiating.

5.7 Show that the radiant intensity generated by a fluctuating three-dimensional source distribution $Q(\mathbf{r}, t)$ is given by the expression

$$J_\omega(\mathbf{s}) = (2\pi)^6 \tilde{W}_Q(-k\mathbf{s}, k\mathbf{s}, \omega),$$

where $\tilde{W}_Q(\mathbf{K}_1, \mathbf{K}_2, \omega)$ is the six-dimensional spatial Fourier transform of the cross-spectral density function $W_Q(\mathbf{r}_1, \mathbf{r}_2, \omega) = \langle Q^*(\mathbf{r}_1, \omega)Q(\mathbf{r}_2, \omega) \rangle$ of the source, i.e.

$$\tilde{W}_Q(\mathbf{K}_1, \mathbf{K}_2, \omega) = \frac{1}{(2\pi)^6} \iint W_Q(\mathbf{r}_1, \mathbf{r}_2, \omega) e^{-i(\mathbf{K}_1 \cdot \mathbf{r}_1 + \mathbf{K}_2 \cdot \mathbf{r}_2)} d^3r_1\, d^3r_2.$$

Show also that when the source is spatially completely incoherent the radiant intensity
is independent of direction.

5.8 Consider a completely coherent, statistically stationary, three-dimensional source distribution $Q(\mathbf{r}, t)$ occupying a finite volume. The cross-spectral density of such a source
has the factorized form

$$W_Q(\mathbf{r}_1', \mathbf{r}_2', \omega) = G^*(\mathbf{r}_1', \omega)G(\mathbf{r}_2', \omega).$$

(a) With the help of the result stated in the previous problem, find an expression for
the radiant intensity of the field generated by the source.
(b) Suppose that the source is uniform, equi-phasal and spherical. Show that for certain values of its radius the source will not radiate.

6

Coherence effects in scattering

When light is incident on an object, it deviates from its original path, i.e. it is scattered by the object. There are many different types of scattering, for example, scattering by atoms or molecules, by dust particles, or by macroscopic bodies. The bodies may be homogeneous or inhomogeneous, isotropic or anisotropic and their behavior may be changing in time. One then speaks of *static* or *dynamic scattering*, respectively. The response of the scatterer may be linear or non-linear and the medium may be deterministic or random.

It is clear from these remarks that light scattering is a very broad subject. In this chapter we will consider only one class of scattering process, albeit a rather broad one: namely scattering on a linear, isotropic, statistically stationary medium. We will discuss both deterministic and stochastic fields incident on a static scattering medium, both when the medium is deterministic and when it is stochastic.

6.1 Scattering of a monochromatic plane wave on a deterministic medium

Let us first consider the scattering of a monochromatic wave

$$V^{(i)}(\mathbf{r}, t) = U^{(i)}(\mathbf{r}, \omega)e^{-i\omega t}, \tag{1}$$

incident upon a linear scatterer, which occupies a finite domain D in free space (Fig. 6.1). Here $\mathbf{r}$ denotes the position vector of any point either outside or inside the scatterer, t denotes the time and ω denotes the frequency. We assume that the physical properties of the medium are characterized by a refractive index $n(\mathbf{r}, \omega)$.

Let

$$V(\mathbf{r}, t) = U(\mathbf{r}, \omega)e^{-i\omega t} \tag{2}$$

be the total field at a point $\mathbf{r}$. $U(\mathbf{r}, \omega)$ then satisfies the equation

$$\nabla^2 U(\mathbf{r}, \omega) + k^2 n^2(\mathbf{r}, \omega)U(\mathbf{r}, \omega) = 0, \tag{3}$$

where k is the free-space wave number associated with frequency ω, i.e.

$$k = \omega/c, \tag{4}$$

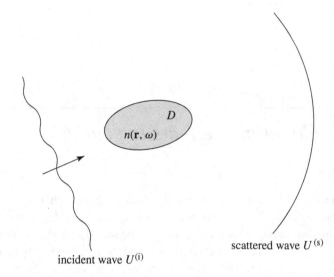

incident wave $U^{(i)}$

scattered wave $U^{(s)}$

Fig. 6.1 Illustrating the notation relating to scattering.

c being the speed of light in vacuum. It is convenient to rewrite Eq. (3) in the form

$$\nabla^2 U(\mathbf{r}, \omega) + k^2 U(\mathbf{r}, \omega) = -4\pi F(\mathbf{r}, \omega) U(\mathbf{r}, \omega), \qquad (5)$$

where the quantity

$$F(\mathbf{r}, \omega) = \frac{1}{4\pi} k^2 [n^2(\mathbf{r}, \omega) - 1] \qquad (6)$$

is called the *scattering potential* of the medium. Because of the relation

$$n^2(\mathbf{r}, \omega) = 1 + 4\pi\eta(\mathbf{r}, \omega) \qquad (7)$$

between the refractive index $n(\mathbf{r}, \omega)$ and the dielectric susceptibility $\eta(\mathbf{r}, \omega)$, the scattering potential may be expressed in the simple form

$$F(\mathbf{r}, \omega) = k^2 \eta(\mathbf{r}, \omega). \qquad (8)$$

Let us represent the field $U(\mathbf{r}, \omega)$ as the sum of the incident field $U^{(i)}(\mathbf{r}, \omega)$ and the scattered field $U^{(s)}(\mathbf{r}, \omega)$, viz.,

$$U(\mathbf{r}, \omega) = U^{(i)}(\mathbf{r}, \omega) + U^{(s)}(\mathbf{r}, \omega). \qquad (9)$$

This relation may, in fact, be regarded as defining the scattered field $U^{(s)}(\mathbf{r}, \omega)$. The incident field will be assumed to satisfy the Helmholtz equation

$$(\nabla^2 + k^2) U^{(i)}(\mathbf{r}, \omega) = 0 \qquad (10)$$

throughout all space.

By the use of well-known vector identities one can show from Eqs. (5) and (10), when Eq. (9) is also used, that, with the scattered field $U^{(s)}(\mathbf{r}, \omega)$ assumed to behave at infinity as an outgoing spherical wave, the total field obeys the equation (cf. B&W, Section 3.1.1)

$$U(\mathbf{r}, \omega) = U^{(i)}(\mathbf{r}, \omega) + \int_D F(\mathbf{r}', \omega) U(\mathbf{r}', \omega) G(|\mathbf{r} - \mathbf{r}'|, \omega) \mathrm{d}^3 r', \tag{11}$$

where $G(R, \omega)$ $(R = |\mathbf{r} - \mathbf{r}'|)$ is the outgoing free-space Green function

$$G(R, \omega) = e^{ikR}/R \tag{12}$$

of the Helmholtz operator. Equation (11) together with Eq. (9) is the basic integral equation for the scattered field, for scattering of a monochromatic wave on a medium characterized by the scattering potential $F(\mathbf{r}, \omega)$. It is generally known as *the integral equation of potential scattering.*

In general, it is not possible to solve Eq. (11) in closed form. However, when the scattering is weak, a relatively simple approximate closed-form solution of Eq. (11) can readily be deduced. By weak scattering we mean that the magnitude $|U^{(s)}|$ of the scattered field is much smaller than the magnitude $|U^{(i)}|$ of the incident field, i.e. that

$$|U^{(s)}(\mathbf{r}, \omega)| \ll |U^{(i)}(\mathbf{r}, \omega)| \tag{13}$$

throughout the scatterer. It is clear from Eqs. (6) and (11) that the scattering will be weak if the refractive index differs only slightly from unity. In such a case, one can, to a good approximation, replace the total field U in the integral in Eq. (11) by the incident field $U^{(i)}$. The integral equation (11) then becomes

$$U(\mathbf{r}, \omega) \approx U^{(i)}(\mathbf{r}, \omega) + \int_D F(\mathbf{r}', \omega) U^{(i)}(\mathbf{r}', \omega) G(|\mathbf{r} - \mathbf{r}'|, \omega) \mathrm{d}^3 r', \tag{14}$$

which is known as the *first-order Born approximation* to the solution of the integral equation of potential scattering. We note that, unlike Eq. (11), Eq. (14) is not an integral equation but is, in fact, a solution to the scattering problem in terms of the incident field $U^{(i)}$ and the scattering potential $F(\mathbf{r}, \omega)$.

It is of interest to note that within the accuracy of the first-order Born approximation Eq. (5) reduces to

$$\nabla^2 U(\mathbf{r}, \omega) + k^2 U(\mathbf{r}, \omega) = -4\pi F(\mathbf{r}, \omega) U^{(i)}(\mathbf{r}, \omega). \tag{15}$$

This equation is identical to the equation for radiation from a scalar source distribution[1]

$$\rho(\mathbf{r}, \omega) = F(\mathbf{r}, \omega) U^{(i)}(\mathbf{r}, \omega), \tag{16}$$

occupying the volume region D. Hence within the accuracy of the first-order Born approximation the process of scattering on a static, linear medium and the process of radiation from a localized source distribution are mathematically equivalent to each other.

---

[1] See, for example, C. H. Papas, *Theory of Electromagnetic Wave Propagation* (McGraw-Hill, New York, 1965), Eq. (56), p. 11.

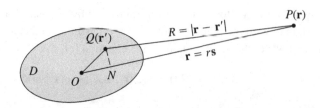

Fig. 6.2 Notation relating to derivation of the far-zone approximation (19) of Section 6.1 to the free-space Green function $G(|\mathbf{r} - \mathbf{r}'|, \omega)$.

For later purposes we derive from Eq. (14) an expression for the scattered field in the far zone of the scattering medium, assuming that the wave incident on the scatterer is a monochromatic plane wave of frequency ω, which propagates in the direction of a real unit vector $\mathbf{s}_0$:

$$U^{(i)}(\mathbf{r}, \omega) = a(\omega)e^{ik\mathbf{s}_0 \cdot \mathbf{r}}. \tag{17}$$

Under these circumstances Eq. (14) becomes

$$U(\mathbf{r}, \omega) \approx a(\omega)e^{ik\mathbf{s}_0 \cdot \mathbf{r}} + a(\omega)\int_D F(\mathbf{r}', \omega)e^{ik\mathbf{s}_0 \cdot \mathbf{r}'}G(|\mathbf{r} - \mathbf{r}'|, \omega)d^3r', \tag{18}$$

where Green's function $G(R, \omega) \equiv G(|\mathbf{r} - \mathbf{r}'|, \omega)$ may now be approximated by the expression

$$G(|\mathbf{r} - \mathbf{r}'|, \omega) \sim \frac{e^{ikr}}{r}e^{-iks \cdot \mathbf{r}'}. \tag{19}$$

The approximation given in Eq. (19) is evident from Fig. 6.2, where $r = OP$ and $|\mathbf{r} - \mathbf{r}'| \approx \overline{QP} \sim \overline{OP} - \overline{ON} \approx r - \mathbf{s} \cdot \mathbf{r}'$, N being the foot of the perpendicular dropped from a point $Q(\mathbf{r}')$ onto the line OP.

If we make use of Eq. (19) in Eq. (14) and use Eq. (17) the field in the far zone is seen to be given by the expression

$$U(r\mathbf{s}, \omega) = a(\omega)\left(e^{iks_0 \cdot \mathbf{r}} + A(\mathbf{s}, \omega)\frac{e^{ikr}}{r} \right), \tag{20}$$

where

$$A(\mathbf{s}, \omega) = \int_D F(\mathbf{r}', \omega)e^{-ik(\mathbf{s}-\mathbf{s}_0) \cdot \mathbf{r}'} d^3r'. \tag{21}$$

In mathematical language, Eq. (20) represents the asymptotic approximation to the far field as $kr \rightarrow \infty$ in the fixed direction $\mathbf{s}$. The amplitude function $A(\mathbf{s}, \omega)$ of the spherical wave on

the right-hand side of Eq. (20) is known as the *scattering amplitude*. If we introduce the
Fourier transform $\tilde{F}(\mathbf{K}, \omega)$ of the scattering potential,

$$\tilde{F}(\mathbf{K},\omega) = \int_D F(\mathbf{r}',\omega)e^{i\mathbf{K}\cdot\mathbf{r}'}\, d^3r', \tag{22}$$

then Eq. (21) implies that

$$A(\mathbf{s},\omega) = \tilde{F}[k(\mathbf{s} - \mathbf{s}_0),\omega]. \tag{23}$$

Equation (23) shows that, within the accuracy of the first-order Born approximation, the
scattering amplitude $A(\mathbf{s}, \omega)$ is equal to a component of the three-dimensional spatial
Fourier transform of the scattering potential, namely the component labeled by the spatial-
frequency vector $\mathbf{K} = k(\mathbf{s} - \mathbf{s}_0)$, $\mathbf{s}_0$ being the unit vector along the direction of incidence.

6.2 Scattering of partially coherent waves on a deterministic medium

We will now consider the more complicated situation when the light incident on a deterministic scatterer is not monochromatic but is partially coherent. The fluctuations of the
light are assumed to be statistically stationary, at least in the wide sense.

Let $W^{(i)}(\mathbf{r}_1, \mathbf{r}_2, \omega)$ be the cross-spectral density function of the incident light. Then as we
learned earlier, $W^{(i)}$ may be expressed in the form [Eq. (13) of Section 4.1]

$$W^{(i)}(\mathbf{r}_1,\mathbf{r}_2,\omega) = \langle U^{(i)*}(\mathbf{r}_1,\omega)U^{(i)}(\mathbf{r}_2,\omega)\rangle, \tag{1}$$

the angular brackets denoting the average over the statistical ensemble of monochromatic
realizations of the incident field.

The scattered field may likewise be represented by an ensemble of monochromatic realizations, $U^{(s)}(\mathbf{r}, \omega)$, whose cross-spectral density function may be represented in a similar
form, viz.,

$$W^{(s)}(\mathbf{r}_1,\mathbf{r}_2,\omega) = \langle U^{(s)*}(\mathbf{r}_1,\omega)U^{(s)}(\mathbf{r}_2,\omega)\rangle. \tag{2}$$

The scattered field $U^{(s)}$ and the incident field $U^{(i)}$ are related by the integral equation of
potential scattering. Within the accuracy of the first-order Born approximation the scattered
field $U^{(s)}$ is given by the integral on the right-hand side of Eq. (14) of Section 6.1. Using
that relation and Eqs. (1) and (2) it follows that

$$W^{(s)}(\mathbf{r}_1,\mathbf{r}_2,\omega) = \int_D\int_D W^{(i)}(\mathbf{r}_1',\mathbf{r}_2',\omega)F^*(\mathbf{r}_1',\omega)F(\mathbf{r}_2',\omega)$$
$$\times\, G^*(|\mathbf{r}_1 - \mathbf{r}_1'|, \omega)G(|\mathbf{r}_2 - \mathbf{r}_2'|, \omega)d^3r_1'\, d^3r_2'. \tag{3}$$

To obtain a clearer insight into some consequences of Eq. (3), let us express the cross-
spectral density $W^{(i)}(\mathbf{r}_1', \mathbf{r}_2', \omega)$ in terms of the spectral densities $S^{(i)}(\mathbf{r}_1', \omega)$ and $S^{(i)}(\mathbf{r}_2', \omega)$ of
the incident light and of its spectral degree of coherence [Eq. (6b) of Section 4.2], viz.,

$$W^{(i)}(\mathbf{r}_1', \mathbf{r}_2', \omega) = \sqrt{S^{(i)}(\mathbf{r}_1', \omega)} \sqrt{S^{(i)}(\mathbf{r}_2', \omega)} \, \mu^{(i)}(\mathbf{r}_1', \mathbf{r}_2', \omega). \tag{4}$$

Equation (3) for the cross-spectral density of the scattered field then becomes

$$W^{(s)}(\mathbf{r}_1, \mathbf{r}_2, \omega) = \int_D \int_D \sqrt{S^{(i)}(\mathbf{r}_1', \omega)} \sqrt{S^{(i)}(\mathbf{r}_2', \omega)} \, \mu^{(i)}(\mathbf{r}_1', \mathbf{r}_2', \omega) F^*(\mathbf{r}_1', \omega) F(\mathbf{r}_2', \omega)$$
$$\times \, G^*(|\mathbf{r}_1 - \mathbf{r}_1'|, \omega) G(|\mathbf{r}_2 - \mathbf{r}_2'|, \omega) \mathrm{d}^3 r_1' \, \mathrm{d}^3 r_2'. \tag{5}$$

If, as is frequently the case, the spectrum of the light incident on the scatterer is independent of position, then we may set $S^{(i)}(\mathbf{r}_1', \omega) = S^{(i)}(\mathbf{r}_2', \omega) \equiv S^{(i)}(\omega)$ in Eq. (5), which then becomes

$$W^{(s)}(\mathbf{r}_1, \mathbf{r}_2, \omega) = S^{(i)}(\omega) \int_D \int_D \mu^{(i)}(\mathbf{r}_1', \mathbf{r}_2', \omega) F^*(\mathbf{r}_1', \omega) F(\mathbf{r}_2', \omega)$$
$$\times \, G^*(|\mathbf{r}_1 - \mathbf{r}_1'|, \omega) G(|\mathbf{r}_2, \mathbf{r}_2'|, \omega) \mathrm{d}^3 r_1' \, \mathrm{d}^3 r_2'. \tag{6}$$

In particular, if the two field points coincide, i.e. if $\mathbf{r}_1 = \mathbf{r}_2 \equiv \mathbf{r}$ say, then the left-hand side of Eq. (6) will represent the spectrum of the scattered field, $S^{(s)}(\mathbf{r}, \omega)$, and the equation reduces to

$$S^{(s)}(\mathbf{r}, \omega) = S^{(i)}(\omega) \int_D \int_D \mu^{(i)}(\mathbf{r}_1', \mathbf{r}_2', \omega) F^*(\mathbf{r}_1', \omega) F(\mathbf{r}_2', \omega)$$
$$\times \, G^*(|\mathbf{r} - \mathbf{r}_1'|, \omega) G(|\mathbf{r} - \mathbf{r}_2'|, \omega) \mathrm{d}^3 r_1' \, \mathrm{d}^3 r_2' \tag{7}$$

This formula shows that, even on static scattering, the spectrum of the scattered field differs, in general, from the spectrum of the incident field, the change arising from (1) the spatial coherence properties of the incident light, (2) the frequency dependence of the potential (due to dispersion of the medium) and (3) the frequency dependence of the free-space Green function. Usually, with narrow-band incident light, the change will be caused mainly by the spatial-coherence properties of the incident light, which are characterized by the spectral degree of coherence $\mu^{(i)}(\mathbf{r}_1', \mathbf{r}_2', \omega)$.

We noted earlier [see the remark that follows Eq. (15) of Section 6.1] that, within the accuracy of the first Born approximation, the processes of radiation and of scattering are equivalent to each other. We also learned in Section 5.6 that the spectrum of a field radiated by a partially coherent source may differ from the spectrum of the source. Hence the result that we just derived, namely that the spectrum of the light scattered on a time-independent medium (i.e. for static scattering) will, in general, differ from the spectrum of the incident light, was to be expected.

We will illustrate some of the preceding results by a simple example. Suppose that a polychromatic plane wave, propagating in the direction of a unit vector $\mathbf{s}_0$, is incident on the scatterer. We may represent the wave by a statistical ensemble (denoted by curly brackets)

$$\{U^{(i)}(\mathbf{r}, \omega)\} = \{a(\omega)\}e^{ik\mathbf{s}_0 \cdot \mathbf{r}}, \tag{8}$$

$a(\omega)$ being a random amplitude. On substituting from Eq. (8) into Eq. (1) we obtain for the cross-spectral density function of the wave incident on the scatterer the expression

$$W^{(i)}(\mathbf{r}_1, \mathbf{r}_2, \omega) = S^{(i)}(\omega)e^{iks_0 \cdot (\mathbf{r}_2 - \mathbf{r}_1)}, \tag{9}$$

where

$$S^{(i)}(\omega) = \langle a^*(\omega)a(\omega) \rangle \tag{10}$$

represents the spectrum of the incident wave. It follows at once from Eq. (9) and from Eq. (4) that the spectral degree of coherence of the incident wave is given by the expression

$$\mu^{(i)}(\mathbf{r}_1, \mathbf{r}_2, \omega) = e^{iks_0 \cdot (\mathbf{r}_2 - \mathbf{r}_1)}. \tag{11}$$

Since in this case $|\mu^{(i)}(\mathbf{r}_1, \mathbf{r}_2, \omega)| = 1$ for all pairs of points $\mathbf{r}_1$ and $\mathbf{r}_2$, the incident wave is spatially completely coherent at the frequency ω throughout all space.

It follows from Eqs. (5) and (11) and from the fact that the spectrum of the incident light is now independent of position that the cross-spectral density of the scattered light is, within the accuracy of the first-order Born approximation, given by the expression

$$W^{(s)}(\mathbf{r}_1, \mathbf{r}_2, \omega) = \langle U^{(s)*}(\mathbf{r}_1, \omega)U^{(s)}(\mathbf{r}_2, \omega) \rangle, \tag{12}$$

where

$$U^{(s)}(\mathbf{r}, \omega) = \sqrt{S^{(i)}(\omega)} \int_D F(\mathbf{r}', \omega)G(|\mathbf{r} - \mathbf{r}'|, \omega)e^{iks_0 \cdot \mathbf{r}'} \, d^3r'. \tag{13}$$

Since the cross-spectral density $W^{(s)}(\mathbf{r}_1, \mathbf{r}_2, \omega)$ is now the product of a function of $\mathbf{r}_1$ and a function of $\mathbf{r}_2$, it follows at once that the spectral degree of coherence $\mu^{(s)}(\mathbf{r}_1, \mathbf{r}_2, \omega)$ of the scattered field, which may be defined by a formula of the form of Eq. (4), is unimodular, implying that the scattered field is also spatially completely coherent at frequency ω. The fact that in the present case the state of coherence has not changed on scattering is rather exceptional. In general not only the spectrum of the light but also its state of coherence will change on scattering.

One is frequently interested only in the behavior of the scattered field in the far zone. One may readily derive an expression for $U^{(s)}(\mathbf{r}, \omega)$ by making use of the far-zone approximation [Eq. (19) of Section 6.1] for the free-space Green function. Equation (13) then becomes

$$\begin{aligned} U^{(s)}(r\mathbf{s}, \omega) &= \sqrt{S^{(i)}(\omega)} \frac{e^{ikr}}{r} \int_D F(\mathbf{r}', \omega)e^{-ik(\mathbf{s}-\mathbf{s}_0) \cdot \mathbf{r}'} \, d^3r' \\ &= \sqrt{S^{(i)}(\omega)}\tilde{F}[k(\mathbf{s} - \mathbf{s}_0), \omega] \frac{e^{ikr}}{r}, \end{aligned} \tag{14}$$

where $\tilde{F}(\mathbf{K}, \omega)$ is the three-dimensional Fourier transform of the scattering potential, defined by Eq. (22) of Section 6.1. From Eqs. (14) and (12) it follows that the cross-spectral density of the scattered field and the spectral density in the far zone are given by the expressions

$$W^{(s)}(r\mathbf{s}_1, r\mathbf{s}_2, \omega) = \frac{1}{r^2} S^{(i)}(\omega)\tilde{F}^*[k(\mathbf{s}_1 - \mathbf{s}_0), \omega]\tilde{F}[k(\mathbf{s}_2 - \mathbf{s}_0), \omega] \tag{15}$$

and

$$S^{(s)}(r\mathbf{s}, \omega) \equiv W^{(s)}(r\mathbf{s}, r\mathbf{s}, \omega) = \frac{1}{r^2} S^{(i)}(\omega)\left|\tilde{F}[k(\mathbf{s} - \mathbf{s}_0), \omega]\right|^2. \tag{16}$$

The preceding formulas refer to coherence properties in the space–frequency domain. Let us now briefly consider coherence properties in the space–time domain. As we learned earlier, these properties are characterized by the mutual coherence function $\Gamma(\mathbf{r}_1, \mathbf{r}_2, \tau)$, which is the Fourier frequency transform of the cross-spectral density [Eq. (2) of Section 4.1]. In the present case, when the wave incident on the scatterer is a polychromatic plane wave, its cross-spectral density function is given by Eq. (9) and hence

$$\Gamma^{(i)}(\mathbf{r}_1, \mathbf{r}_2, \tau) = \int_0^\infty S^{(i)}(\omega)e^{ik\mathbf{s}_0 \cdot (\mathbf{r}_2 - \mathbf{r}_1)}e^{-i\omega\tau}\, d\omega$$
$$= \int_0^\infty S^{(i)}(\omega)e^{-i\omega[\tau - \mathbf{s}_0 \cdot (\mathbf{r}_2 - \mathbf{r}_1)/c]}\, d\omega. \tag{17}$$

The degree of coherence $\gamma^{(i)}(\mathbf{r}_1, \mathbf{r}_2, \tau)$ of the wave is obtained by normalizing Eq. (17) in accordance with Eq. (10) of Section 3.1 and one finds that

$$\gamma^{(i)}(\mathbf{r}_1, \mathbf{r}_2, \tau) = \frac{\int_0^\infty S^{(i)}(\omega)e^{-i\omega[\tau - \mathbf{s}_0 \cdot (\mathbf{r}_2 - \mathbf{r}_1)/c]}\, d\omega}{\int_0^\infty S^{(i)}(\omega)\, d\omega}. \tag{18}$$

The mutual coherence function of the scattered field in the far zone is obtained by taking the Fourier transform of Eq. (15), which gives

$$\Gamma^{(s)}(r\mathbf{s}_1, r\mathbf{s}_2, \tau) = \frac{1}{r^2}\int_0^\infty S^{(i)}(\omega)\,\tilde{F}^*[k(\mathbf{s}_1 - \mathbf{s}_0), \omega]\,\tilde{F}[k(\mathbf{s}_2 - \mathbf{s}_0), \omega]e^{-i\omega\tau}\, d\omega. \tag{19}$$

An expression for the degree of coherence $\gamma^{(s)}(r\mathbf{s}_1, r\mathbf{s}_2, \tau)$ of the scattered field is obtained by normalizing Eq. (19) in accordance with Eq. (10) of Section 3.1, as above.

6.3 Scattering on random media

6.3.1 General formulas

So far we have assumed that the scatterer is deterministic. The scattering potential $F(\mathbf{r}, \omega)$ is then a well-defined function of position. Frequently, however, this is not so; the scattering potential is then a random function of position. An example of such a situation is the turbulent atmosphere in which the refractive index varies randomly both in time and in space, due to irregular fluctuations in temperature and pressure. Over sufficiently short intervals of time, typically of the order of a tenth of a second, the temporal fluctuations may be neglected and one then speaks of a "frozen model" of the turbulent atmosphere. Under these circumstances one is essentially dealing with static scattering, to which our considerations apply. For such situations an expression for the cross-spectral density of the scattered field that is valid to within the accuracy of the first-order Born approximation is obtained by averaging Eqs. (3) and (5) of the preceding section over the ensemble of the scattering medium. We will denote this average by angular brackets with subscript m (i.e. we will write it as $\langle \ldots \rangle_m$). We then obtain, from Eq. (5) of the preceding section, the formula

$$
W^{(s)}(\mathbf{r}_1, \mathbf{r}_2, \omega) = \int_D \int_D \sqrt{S^{(i)}(\mathbf{r}_1', \omega)} \sqrt{S^{(i)}(\mathbf{r}_2', \omega)} \mu^{(i)}(\mathbf{r}_1', \mathbf{r}_2', \omega) C_F(\mathbf{r}_1', \mathbf{r}_2', \omega)
$$
$$
\times G^*(|\mathbf{r}_1 - \mathbf{r}_1'|, \omega) G(|\mathbf{r}_2 - \mathbf{r}_2'|, \omega) \, d^3 r_1' \, d^3 r_2', \tag{1}
$$

where

$$
C_F(\mathbf{r}_1', \mathbf{r}_2', \omega) = \langle F^*(\mathbf{r}_1', \omega) F(\mathbf{r}_2', \omega) \rangle_m \tag{2}
$$

is the *correlation function of the scattering potential*. It should be noted that Eq. (1) contains implicitly two different averages, namely one taken over the ensemble of the incident field and the other over the ensemble of the scatterer. Since we have assumed that the scattering is weak (in the sense of the first-order Born approximation) we can treat the two averaging processes independently of each other.

The spectral density of the scattered field is obtained at once from Eq. (1) by setting $\mathbf{r}_1 = \mathbf{r}_2 = \mathbf{r}$ and we then find that

$$
S^{(s)}(\mathbf{r}, \omega) = \int_D \int_D \sqrt{S^{(i)}(\mathbf{r}_1', \omega)} \sqrt{S^{(i)}(\mathbf{r}_2', \omega)} \mu^{(i)}(\mathbf{r}_1', \mathbf{r}_2', \omega) C_F(\mathbf{r}_1', \mathbf{r}_2', \omega)
$$
$$
\times G^*(|\mathbf{r} - \mathbf{r}_1'|, \omega) G(|\mathbf{r} - \mathbf{r}_2'|, \omega) d^3 r_1' \, d^3 r_2'. \tag{3}
$$

Although Eqs. (2) and (3) involve averages over macroscopically similar but microscopically different scatterers, one can often deduce the values of these averages, at least to a good approximation, from experiments involving only a single scatterer. For example, the necessarily finite size of the detector aperture will frequently provide spatial averaging, which is essentially equivalent to ensemble averaging.[1]

---

[1] See L. G. Shirley and N. George, *Appl. Opt.* **27** (1988), 1850–1861, Section II, and J. Goodman's contribution in *Laser Speckle and Related Phenomena*, second edition, J. C. Dainty ed. (Springer, New York, 1984), Section 2.6.1.

Expressions for the cross-spectral density and for the spectrum of the scattered field in the far zone may be obtained at once from Eqs. (1) and (3) by again using the far-zone approximation (19) of Section 6.1 for the free-space Green function. Equations (1) and (3) then give

$$W^{(s)}(r\mathbf{s}_1, r\mathbf{s}_2, \omega) \approx \frac{1}{r^2} \int_D \int_D \sqrt{S^{(i)}(\mathbf{r}_1', \omega)} \sqrt{S^{(i)}(\mathbf{r}_2', \omega)} \mu^{(i)}(\mathbf{r}_1', \mathbf{r}_2', \omega) C_F(\mathbf{r}_1', \mathbf{r}_2', \omega)$$
$$\times e^{-ik(\mathbf{s}_2 \cdot \mathbf{r}_2' - \mathbf{s}_1 \cdot \mathbf{r}_1')} \, d^3 r_1' \, d^3 r_2' \tag{4}$$

and

$$S^{(s)}(r\mathbf{s}_1, \omega) \sim \frac{1}{r^2} \int_D \int_D \sqrt{S^{(i)}(\mathbf{r}_1', \omega)} \sqrt{S^{(i)}(\mathbf{r}_2', \omega)} \mu^{(i)}(\mathbf{r}_1', \mathbf{r}_2', \omega) C_F(\mathbf{r}_1', \mathbf{r}_2', \omega)$$
$$\times e^{-iks \cdot (\mathbf{r}_2' - \mathbf{r}_1')} \, d^3 r_1' \, d^3 r_2' \tag{5}$$

as $kr \to \infty$, with the directions $\mathbf{s}_1$, $\mathbf{s}_2$ and $\mathbf{s}$ being kept fixed. An expression for the spectral degree of coherence of the scattered field in the far zone is obtained on substituting from Eqs. (4) and (5) into the formula (6b) of Section 4.2 which defines the spectral degree of coherence, viz.,

$$\mu^{(s)}(r\mathbf{s}_1, r\mathbf{s}_2, \omega) = \frac{W^{(s)}(r\mathbf{s}_1, r\mathbf{s}_2, \omega)}{\sqrt{S^{(s)}(r\mathbf{s}_1, \omega)} \sqrt{S^{(s)}(r\mathbf{s}_2, \omega)}}. \tag{6}$$

We have assumed so far that the scatterer is a continuous medium. The analysis may readily be extended to scattering by a system of particles, a case of considerable practical interest. If the scattering potential of each particle is the same, say $f(\mathbf{r}, \omega)$ and the particles are located at points with position vectors $\mathbf{r}_1, \mathbf{r}_2, \ldots$, the scattering potential of the whole system of particles is

$$F(\mathbf{r}, \omega) = \sum_n f(\mathbf{r} - \mathbf{r}_n, \omega). \tag{7}$$

The correlation function of this scattering potential is evidently

$$C_F(\mathbf{r}_1, \mathbf{r}_2, \omega) = \langle F^*(\mathbf{r}_1, \omega) F(\mathbf{r}_2, \omega) \rangle = \sum_m \sum_n \langle f^*(\mathbf{r}_1 - \mathbf{r}_m, \omega) f(\mathbf{r}_2 - \mathbf{r}_n, \omega) \rangle, \tag{8}$$

where the expectation value is taken over the ensemble of the particles.

The formulas that we have derived in this section and in the preceding one may be used to elucidate many features of the scattering of the light of any state of coherence, on a deterministic or on a random medium. We will now illustrate them by some examples.

6.3.2 Examples

When the incident field is a spatially coherent, polychromatic plane wave, its cross-spectral density is given by Eq. (9) of Section 6.2. On substituting that expression into formula (4), we obtain for the cross-spectral density function of the scattered wave in the far zone the expression

$$W^{(s)}(r\mathbf{s}_1, r\mathbf{s}_2, \omega) \sim \frac{1}{r^2} S^{(i)}(\omega) \int_D \int_D e^{ik\mathbf{s}_0 \cdot (\mathbf{r}_2' - \mathbf{r}_1')} C_F(\mathbf{r}_1', \mathbf{r}_2', \omega) e^{-ik(\mathbf{s}_2 \cdot \mathbf{r}_2' - \mathbf{s}_1 \cdot \mathbf{r}_1')} d^3 r_1' d^3 r_2'$$

$$= \frac{1}{r^2} S^{(i)}(\omega) \tilde{C}_F[-k(\mathbf{s}_1 - \mathbf{s}_0), k(\mathbf{s}_2 - \mathbf{s}_0), \omega], \tag{9}$$

where

$$\tilde{C}_F(\mathbf{K}_1, \mathbf{K}_2, \omega) = \int_D \int_D C_F(\mathbf{r}_1', \mathbf{r}_2', \omega) e^{-i(\mathbf{K}_1 \cdot \mathbf{r}_1' + \mathbf{K}_2 \cdot \mathbf{r}_2')} d^3 r_1' d^3 r_2' \tag{10}$$

is the six-dimensional spatial Fourier transform of the correlation function $C_F(\mathbf{r}_1', \mathbf{r}_2', \omega)$ of the scattering potential.

The spectrum of the scattered field in the far zone of the scatterer is obtained at once from Eq. (9) on setting $\mathbf{s}_1 = \mathbf{s}_2 = \mathbf{s}$. One then finds that

$$S^{(s)}(r\mathbf{s}, \omega) \sim \frac{1}{r^2} S^{(i)}(\omega) \tilde{C}_F[-k(\mathbf{s} - \mathbf{s}_0), k(\mathbf{s} - \mathbf{s}_0), \omega]. \tag{11}$$

Suppose that the scatterer is a Gaussian-correlated, homogeneous isotropic medium. Its correlation function then has the form

$$C_F(\mathbf{r}_1, \mathbf{r}_2, \omega) = \frac{A}{\left(\sigma\sqrt{(2\pi)}\right)^3} e^{-|\mathbf{r}_2 - \mathbf{r}_1|^2/(2\sigma^2)}, \tag{12}$$

where A and σ are positive constants. We assume that σ is small compared with the linear dimensions of the scatterer. Using Eqs. (10) – (12) one finds after a straightforward calculation that the spectral density of the scattered field in the far zone is given by the expression

$$S^{(s)}(r\mathbf{s}, \omega) = \frac{AV}{r^2} S^{(i)}(\omega) e^{-2(k\sigma)^2 \sin^2(\theta/2)}, \tag{13}$$

where V is the scattering volume and θ is the angle of scattering ($\mathbf{s} \cdot \mathbf{s}_0 = \cos\theta$).

Let us compare this situation with the case when the incident light is ambient, rather than spatially coherent. The cross-spectral density function of the incident field has the form

$$W^{(i)}(\mathbf{r}_1, \mathbf{r}_2, \omega) = S^{(i)}(\omega) \frac{\sin(k|\mathbf{r}_2 - \mathbf{r}_1|)}{|\mathbf{r}_2 - \mathbf{r}_1|} \tag{14}$$

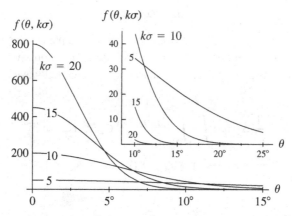

Fig. 6.3 The ratio $f(\theta; k\sigma) = [S^{(s)}(\omega)]_{coh}/[S^{(s)}]_{amb}$ of the spectra of a spatially coherent plane light wave and of ambient light, scattered on a Gaussian-correlated isotropic medium of r.m.s. width σ, calculated to within the accuracy of the first Born approximation. [Adapted from J. Jannson, T. Jannson and E. Wolf, *Opt. Lett.* **13** (1988), 1060–1062.]

rather than the form given by Eq. (9) of Section 6.2. In this case, Eq. (5) may be shown, after some calculations, to give

$$S^{(s)}(r\mathbf{s},\,\omega) = \frac{AV}{r^2}\frac{1}{2(k\sigma)^2}\,S^{(i)}(\omega)\left[1 - e^{-2(k\sigma)^2}\right]. \tag{15}$$

One may readily verify that when $k\sigma \ll 1$ (highly uncorrelated scatterer) Eqs. (15) and (13) become identical. However, when $k\sigma \gg 1$, i.e. when the correlation length σ of the scatterer is much greater than the reduced wavelength $\lambda/(2\pi) = 1/k = c/\omega$ of the incident light, the angular distributions of the scattered light in the far zone in the two cases are completely different. This fact is illustrated in Fig. 6.3.

On returning to Eq. (11) we see, as was to be expected from our earlier remarks relating to spectral changes produced on scattering and from the comment following Eq. (16) in Section 6.1, that the spectrum of the scattered field in the far zone will, in general, differ from the source spectrum.[1]

Effects of spectral coherence on the angular distribution of the scattered light were studied experimentally[2] and confirmed some of the main features relating to effects of the state of coherence of the incident light on the angular distribution of the scattered radiation.

As another example, let us consider the scattering of a spatially coherent polychromatic plane wave on a random distribution of identical particles. The spectral density of the scattered

[1] See, for example, E. Wolf, J. T. Foley and F. Gori, *J. Opt. Soc. Amer.* **A6** (1989), 1142–1149; errata, *Ibid.* **A7** (1990), 173.
[2] F. Gori, C. Palma and M. Santersiero, *Opt. Commun.* **74** (1990), 353–356.

light in the far zone is given by Eq. (11), with $\tilde{C}_F$ defined by Eq. (10). The quantity C_F in this case, is given by the formula

$$\tilde{C}_F(-\mathbf{K}, \mathbf{K}, \omega) = \sum_m \sum_n \int_D \int_D \langle f^*(\mathbf{r}_1' - \mathbf{r}_m, \omega) f(\mathbf{r}_2' - \mathbf{r}_n, \omega) \rangle e^{-i[\mathbf{K}\cdot(\mathbf{r}_2' - \mathbf{r}_1')]} \, d^3 r_1' \, d^3 r_2'.$$

(16)

Let us introduce new variables $\mathbf{R}_{1m} = \mathbf{r}_1' - \mathbf{r}_m$, $\mathbf{R}_{2n} = \mathbf{r}_2' - \mathbf{r}_n$. The formula (16) then becomes

$$\tilde{C}_F(-\mathbf{K}, \mathbf{K}, \omega) = \left\langle \sum_m \sum_n e^{-i\mathbf{K}\cdot(\mathbf{r}_n - \mathbf{r}_m)} \right\rangle$$
$$\times \int_D \int_D \langle f^*(\mathbf{R}_{1m}, \omega) f(\mathbf{R}_{2n}, \omega) \rangle e^{-i\mathbf{K}\cdot(\mathbf{R}_{2n} - \mathbf{R}_{1m})} \, d^3 R_{1m} \, d^3 R_{2n},$$

(17)

which may be expressed in the compact form

$$\tilde{C}_F(-\mathbf{K}, \mathbf{K}, \omega) = |\tilde{f}(\mathbf{K}, \omega)|^2 S(\mathbf{K}),$$

(18)

where $\tilde{f}(\mathbf{K}, \omega)$ is the Fourier transform of $f(\mathbf{r}, \omega)$ and

$$S(\mathbf{K}) = \left\langle \left| \sum_m e^{-i\mathbf{K}\cdot\mathbf{r}_m} \right|^2 \right\rangle$$

(19)

is the so-called *generalized structure function of the particle system*.[1]

On substituting from Eq. (18) into Eq. (11) we obtain for the spectrum of the scattered field in the far zone the expression

$$S^{(s)}(r\mathbf{s}, \omega) = \frac{1}{r^2} S^{(i)}(\omega) \left| \tilde{f}[k(\mathbf{s} - \mathbf{s}_0), \omega] \right|^2 S[k(\mathbf{s} - \mathbf{s}_0)].$$

(20)

This formula shows that the spectrum of the incident wave is changed by scattering on a system of particles, the change being caused by the scattering potential $f(\mathbf{r}, \omega)$ of the individual particles and by the structure function $S(\mathbf{K})$ of the particle system.

Some examples of the changes in the spectrum produced by particle scattering were discussed by Dogariu and Wolf.[2] The possibility of utilizing spectral changes generated by scattering to determine density correlation functions in some particle systems was also noted.[3]

6.3.3 Scattering on a quasi-homogeneous medium[4]

An important class of random media is constituted by the so-called *quasi-homogeneous media* (also called *locally homogeneous media*).[5] To explain the basic properties of such

[1] This function is proportional to the structure factor, which plays an important role in the theory of disordered systems. [See, for example, J. M. Ziman, *Models of Disorder* (Cambridge University Press, Cambridge, 1979)].
[2] A. Dogariu and E. Wolf, *Opt. Lett.* **23** (1998), 1340–1342.
[3] G. Gbur and E. Wolf, *Opt. Commun.* **168** (1999), 39–45.
[4] The analysis in this section is largely based on the following papers: W. H. Carter and E. Wolf, *Opt. Commun.* **67** (1988), 85–90; D. G. Fischer and E. Wolf, *J. Opt. Soc. Amer.* **A11** (1994), 1128–1135; and T. D. Visser, D. G. Fischer and E. Wolf, *J. Opt. Soc. Amer.* **A23** (2006), 1631–1638.
[5] Scattering from such media appears to have been first considered by R. A. Silverman, *Proc. Cambridge Philos. Soc.* **54** (1958), 530–537.

media, we first introduce the normalized correlation coefficient of the scattering potential, defined by the formula

$$\mu_F(\mathbf{r}_1, \mathbf{r}_2, \omega) = \frac{C_F(\mathbf{r}_1, \mathbf{r}_2, \omega)}{\sqrt{C_F(\mathbf{r}_1, \mathbf{r}_1, \omega)}\sqrt{C_F(\mathbf{r}_2, \mathbf{r}_2, \omega)}}, \tag{21}$$

where $C_F(\mathbf{r}_1, \mathbf{r}_2, \omega)$ is the correlation function, defined by Eq. (2), of the scattering potential $F(\mathbf{r}, \omega)$. It is useful to introduce also the function

$$I_F(\mathbf{r}, \omega) \equiv C_F(\mathbf{r}, \mathbf{r}, \omega) = \langle F^*(\mathbf{r}, \omega)F(\mathbf{r}, \omega)\rangle_m, \tag{22}$$

which is a measure of the strength of the scattering potential at the point $\mathbf{r}$.

A quasi-homogeneous medium is characterized by the properties that at each frequency ω

(1) the normalized correlation coefficient $\mu_F(\mathbf{r}_1, \mathbf{r}_2, \omega)$ depends on the two spatial variables $\mathbf{r}_1$ and $\mathbf{r}_2$ only through the difference $\mathbf{r}_2 - \mathbf{r}_1$, in which case we will write $\mu_F(\mathbf{r}_2 - \mathbf{r}_1, \omega)$ in place of $\mu_F(\mathbf{r}_1, \mathbf{r}_2, \omega)$; and
(2) the strength $I_F(\mathbf{r}, \omega)$ of the scattering potential varies much more slowly with $\mathbf{r}$ than $\mu_F(\mathbf{r}_2 - \mathbf{r}_1, \omega)$ varies with the difference $\mathbf{r}' = \mathbf{r}_2 - \mathbf{r}_1$. Hence $I_F(\mathbf{r}, \omega)$ remains nearly constant over distances for which $|\mu_F(\mathbf{r}', \omega)|$ has an appreciable value.

These properties are sometimes expressed by saying that $I_F(\mathbf{r}, \omega)$ is a *slow* function of $\mathbf{r}$ and that $\mu_F(\mathbf{r}', \omega)$ is a *fast* function of $\mathbf{r}'$. Examples of media with such properties are the troposphere and confined plasmas.

The conditions (1) and (2) are strictly analogous to conditions that characterize quasi-homogeneous sources, as we discussed in Section 5.3.2. We showed there[1] that such sources and the radiation generated by them obey certain reciprocity relations. Because of the previously mentioned analogy between radiation and scattering, we can expect that somewhat similar results will apply in connection with scattering. We will now show that this is indeed the case.

Let us rewrite Eq. (21) in the form

$$C_F(\mathbf{r}_1, \mathbf{r}_2, \omega) = \sqrt{I_F(\mathbf{r}_1, \omega)}\sqrt{I_F(\mathbf{r}_2, \omega)}\mu_F(\mathbf{r}_1, \mathbf{r}_2, \omega). \tag{23}$$

Because of the properties which characterize quasi-homogeneous scatterers, C_F may be approximated by the formula

$$C_F(\mathbf{r}_1, \mathbf{r}_2, \omega) = I_F\left(\frac{\mathbf{r}_1 + \mathbf{r}_2}{2}, \omega\right)\mu_F(\mathbf{r}_2 - \mathbf{r}_1, \omega). \tag{24}$$

Suppose now that the wave incident on such a medium is a polychromatic plane wave of spectral density $S(\omega)$, propagating in a direction specified by a unit vector $\mathbf{s}_0$. According to Eq. (11) of Section 6.2, the spectral degree of coherence of such a wave has the simple form

---

[1] In Section 5.3.2 we considered planar quasi-homogeneous sources. However, it is not difficult to show that strictly similar results apply to three-dimensional sources of this class (see M&W, Section 5.8.2).

$$\mu^{(i)}(\mathbf{r}_1, \mathbf{r}_2, \omega) = e^{iks_0 \cdot (\mathbf{r}_2 - \mathbf{r}_1)}. \tag{25}$$

On using Eq. (25) and the fact that the spectrum of such a wave is independent of position, the formula (9) for the cross-spectral density of the scattered field in the far zone applies. The Fourier transform which enters that formula is now given by the expression

$$\tilde{C}_F(\mathbf{K}_1, \mathbf{K}_2, \omega) = \int_D \int_D I_F\left(\frac{\mathbf{r}_1' + \mathbf{r}_2'}{2}, \omega\right) \mu_F(\mathbf{r}_2' - \mathbf{r}_1', \omega) e^{-i(\mathbf{K}_1 \cdot \mathbf{r}_1' + \mathbf{K}_2 \cdot \mathbf{r}_2')} \, d^3 r_1' \, d^3 r_2'. \tag{26}$$

The integrations on the right can be considerably simplified by changing the variables of integration. Let us set

$$\mathbf{r} = \tfrac{1}{2}(\mathbf{r}_1' + \mathbf{r}_2'), \qquad \mathbf{r}' = \mathbf{r}_2' - \mathbf{r}_1'. \tag{27a}$$

Then we have the inverse relations

$$\mathbf{r}_1' = \mathbf{r} - \tfrac{1}{2}\mathbf{r}', \qquad \mathbf{r}_2' = \mathbf{r} + \tfrac{1}{2}\mathbf{r}', \tag{27b}$$

and Eq. (26) becomes, after some calculations,

$$\tilde{C}_F(\mathbf{K}_1, \mathbf{K}_2, \omega) = \int_D \int_D I_F(\mathbf{r}, \omega) \mu_F(\mathbf{r}', \omega) e^{-i[\mathbf{K}_1 \cdot (\mathbf{r} - \frac{1}{2}\mathbf{r}') + \mathbf{K}_2 \cdot (\mathbf{r} + \frac{1}{2}\mathbf{r}')]} \, d^3 r \, d^3 r', \tag{28}$$

where we have used the fact that the Jacobian of the transformation (27) is unity, as may readily be verified.

From Eq. (28) one finds that

$$\tilde{C}_F(\mathbf{K}_1, \mathbf{K}_2, \omega) = \tilde{I}_F(\mathbf{K}_1 + \mathbf{K}_2, \omega) \tilde{\mu}_F\left(\tfrac{1}{2}(\mathbf{K}_2 - \mathbf{K}_1), \omega\right), \tag{29}$$

where

$$\tilde{I}_F(\mathbf{K}, \omega) = \int_D I_F(\mathbf{r}, \omega) e^{-i\mathbf{K} \cdot \mathbf{r}} \, d^3 r \tag{30a}$$

and

$$\tilde{\mu}_F(\mathbf{K}', \omega) = \int_D \mu_F(\mathbf{r}', \omega) e^{-i\mathbf{K}' \cdot \mathbf{r}'} \, d^3 r', \tag{30b}$$

are the Fourier transforms of I_F and μ_F respectively. On substituting for $\tilde{C}_F$ from Eq. (29) into Eq. (9), we obtain the following expression for the cross-spectral density of the scattered field in the far zone of a quasi-homogeneous medium illuminated by a polychromatic plane wave:

$$W^{(s)}(r\mathbf{s}_1, r\mathbf{s}_2, \omega) \sim \frac{1}{r^2} S^{(i)}(\omega) \tilde{I}_F[k(\mathbf{s}_2 - \mathbf{s}_1), \omega] \, \tilde{\mu}_F\left[k\left(\frac{\mathbf{s}_1 + \mathbf{s}_2}{2} - \mathbf{s}_0\right), \omega\right]. \tag{31}$$

From Eq. (31) it follows that the spectral density is given by the formula

$$S^{(s)}(r\mathbf{s}, \omega) = \frac{1}{r^2} S^{(i)}(\omega) \tilde{I}_F(0, \omega) \tilde{\mu}_F[k(\mathbf{s} - \mathbf{s}_0), \omega], \tag{32}$$

where, according to Eq. (30a),

$$\tilde{I}_F(0, \omega) = \int_D \langle |F(\mathbf{r}, \omega)|^2 \rangle_m \, \mathrm{d}^3 r'. \tag{33}$$

The spectral degree of coherence of the scattered field can also be readily determined from Eq. (31) and one finds that

$$\mu^{(s)}(r\mathbf{s}_1, r\mathbf{s}_2, \omega) = \frac{\tilde{I}_F[k(\mathbf{s}_2 - \mathbf{s}_1), \omega]}{I_F(0, \omega)} G(\mathbf{s}_1, \mathbf{s}_2; \mathbf{s}_0, \omega), \tag{34}$$

with

$$G(\mathbf{s}_1, \mathbf{s}_2; \mathbf{s}_0, \omega) = \frac{\tilde{\mu}_F\left[k\left(\dfrac{\mathbf{s}_1 + \mathbf{s}_2}{2} - \mathbf{s}_0\right), \omega\right]}{\sqrt{\tilde{\mu}_F[k(\mathbf{s}_1 - \mathbf{s}_0), \omega]} \sqrt{\tilde{\mu}_F[k(\mathbf{s}_2 - \mathbf{s}_0), \omega]}}. \tag{35}$$

Because for a quasi-homogeneous scatterer $\mu_F(\mathbf{r}', \omega)$ is a fast function of its spatial argument, it follows from a well-known reciprocity theorem concerning Fourier-transform pairs that $\tilde{\mu}_F(\mathbf{K}, \omega)$ is a slow function of $\mathbf{K}$. Under these circumstances

$$\tilde{\mu}_F[k(\mathbf{s}_1 - \mathbf{s}_0), \omega] \approx \tilde{\mu}_F[k(\mathbf{s}_2 - \mathbf{s}_0), \omega] \approx \tilde{\mu}_F\left[k\frac{(\mathbf{s}_1 + \mathbf{s}_2)}{2} - \mathbf{s}_0, \omega\right]. \tag{36}$$

Consequently Eq. (34) simplifies to

$$\mu^{(s)}(r\mathbf{s}_1, r\mathbf{s}_2, \omega) \approx \frac{\tilde{I}_F[k(\mathbf{s}_2 - \mathbf{s}_1), \omega]}{\tilde{I}_F(0, \omega)}. \tag{37}$$

Equations (32) and (37) bring into evidence the following two *reciprocity relations* for scattering of a polychromatic plane wave on a quasi-homogeneous medium:

(1) the spectrum of the scattered field in the far zone is proportional to the Fourier transform of the correlation coefficient of the scatterer; and

(2) the spectral degree of coherence of the scattered field in the far zone is proportional to the Fourier transform of the strength of the scattering potential.

These reciprocity relations for scattering are analogues to the reciprocity relations for radiation from quasi-homogeneous sources, which we encountered in Section 5.3.2 [Eqs. (16) and (21) of that section].

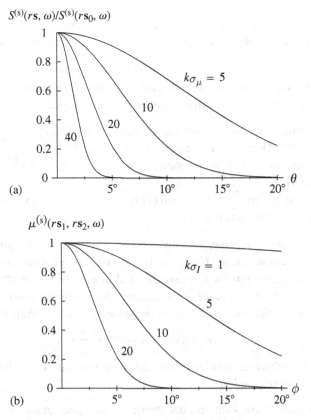

$S^{(s)}(r\mathbf{s}, \omega)/S^{(s)}(r\mathbf{s}_0, \omega)$

$k\sigma_\mu = 5$

(a)

$\mu^{(s)}(r\mathbf{s}_1, r\mathbf{s}_2, \omega)$

$k\sigma_I = 1$

(b)

Fig. 6.4 Illustrating the effect of scattering of a plane polychromatic wave on a quasi-homogeneous, Gaussian-correlated spherical medium. The normalized correlation coefficient μ_F and the strength factor I_F of the scattering potential of the medium are taken to be given by the expressions $\mu_F(\mathbf{r}', \omega) = e^{-r'^2/(2\sigma_\mu^2)}$ and $I_F(\mathbf{r}, \omega) = e^{-r^2/(2\sigma_I^2)}$. (a) The normalized spectral density $S^{(s)}(r\mathbf{s}, \omega)/S^{(s)}(r\mathbf{s}_0, \omega)$ of the scattered field in the far zone, as a function of the angle θ between the direction of scattering $\mathbf{s}$ and the direction of incidence $\mathbf{s}_0$. (b) The spectral degree of coherence $\mu^{(s)}(r\mathbf{s}_1, r\mathbf{s}_2, \omega)$ of the scattered field in the far zone in the directions $\mathbf{s}_1$ and $\mathbf{s}_2$ located symmetrically with respect to the direction $\mathbf{s}_0$ of incidence and $\mathbf{s}_1 \cdot \mathbf{s}_0 = \mathbf{s}_2 \cdot \mathbf{s}_0 = \cos\phi$. [Adapted from T. D. Visser, D. G. Fischer and E. Wolf, *J. Opt. Soc. Amer.* **A23** (2006), 1631–1638.]

Results of numerical calculations relating to scattering from a Gaussian-correlated medium, based on these two reciprocity relations, are shown in Fig. 6.4.

PROBLEMS

6.1 A plane monochromatic wave propagating in free space is scattered by a homogeneous sphere of radius a, whose refractive index is close to unity. Assuming that the effect of the discontinuity of the scattering potential at the boundary of the sphere may be neglected, determine, within the accuracy of the first-order Born approximation, the scattering amplitude in a direction that makes an angle θ with the direction of incidence.

6.2 A polychromatic plane wave is incident on a weak random scatterer which occupies a finite volume. Derive an expression that indicates how the normalized spectrum of the scattered field in the far zone differs from the normalized spectrum of the incident field.

6.3 A polychromatic plane wave is incident on a deterministic scatterer, which occupies a finite domain D. Derive an expression, valid within the accuracy of the first Born approximation, for the degree of coherence $\gamma(r_1 \mathbf{s}_1, r_2 \mathbf{s}_2, \tau)$, $(\mathbf{s}_1^2 = \mathbf{s}_2^2 = 1)$, of the scattered field in the far zone.

Consider also the special cases when (1) $r_1 = r_2$; and (2) $\mathbf{s}_1 = \mathbf{s}_2$. Discuss the physical significance of these two cases.

6.4 Consider scattering of a plane, spatially coherent wave on a weak, homogeneous scatterer, with correlation function $C_F(\mathbf{r}_1, \mathbf{r}_2, \omega) \equiv C_F(\mathbf{r}_2 - \mathbf{r}_1, \omega)$ of the scattering potential.
 (a) Derive an expression for the cross-spectral density of the spectrum and for the spectral degree of coherence of the scattered field in the far zone.
 (b) Consider also the special case when the scatterer is delta-correlated, i.e. when

$$C_F(\mathbf{r}_1, \mathbf{r}_2, \omega) = A(\omega)\delta^{(3)}(\mathbf{r}_2 - \mathbf{r}_1),$$

$\delta^{(3)}$ being a three-dimensional Dirac delta function and $A(\omega)$ a function which is independent of position.

6.5 A coherent plane wave with spectral density $S^{(0)}(\omega)$ propagates into the half-space $z > 0$ and is scattered on a finite random medium, located in that half-space. The correlation function of the scattering potential $F(\mathbf{r}, \omega)$ is $C_F(\mathbf{r}_1, \mathbf{r}_2, \omega)$.

Show that, within the accuracy of the first Born approximation, the spectrum $S^{(s)}(\mathbf{r}, \omega)$ of the scattered field may be expressed in the form

$$S^{(s)}(\mathbf{r}, \omega) = S^{(0)}(\omega)M(\mathbf{r}, \omega);$$

and derive an explicit expression for the "spectral modifier" $M(\mathbf{r}, \omega)$ for this case.

7

Higher-order coherence effects

7.1 Introduction

In the preceding chapters we have been concerned with coherence phenomena which depend on correlations between field variables at two space–time points, say $(\mathbf{r}_1, t_1)$ and $(\mathbf{r}_2, t_2)$. One then speaks of *second-order* coherence phenomena. When the fluctuations are statistically stationary and the polarization features of the field are not taken into account, such phenomena may be characterized by the "space–time" correlation function (the mutual coherence function) [Eq. (6), Section 3.1]

$$\Gamma(\mathbf{r}_1, \mathbf{r}_2, \tau) = \langle V^*(\mathbf{r}_1, t)V(\mathbf{r}_2, t + \tau) \rangle \tag{1}$$

in the space–time domain; or, equivalently, by the space–frequency correlation function, i.e. the cross-spectral density [Eq. (13) of Section 4.1],

$$W(\mathbf{r}_1, \mathbf{r}_2, \omega) = \langle U^*(\mathbf{r}_1, \omega)U(\mathbf{r}_2, \omega) \rangle. \tag{2}$$

The two correlation functions given in Eqs. (1) and (2) form a Fourier-transform pair [Eq. (2) of Section 4.1].

Although the coherence phenomena that have been discussed so far can be described in terms of such correlation functions, there are other coherence phenomena which must be analyzed by using other correlation functions. Because the theory involving them is somewhat complicated, we will consider only correlation phenomena which can be described by the so-called fourth-order correlation functions, which are defined by the expression[1]

$$\Gamma^{(2,2)}(\mathbf{r}_1, t_1; \mathbf{r}_2, t_2; \mathbf{r}_3, t_3; \mathbf{r}_4, t_4) = \langle V^*(\mathbf{r}_1, t_1)V^*(\mathbf{r}_2, t_2)V(\mathbf{r}_3, t_3)V(\mathbf{r}_4, t_4) \rangle. \tag{3}$$

The first superscript on $\Gamma^{(2,2)}$ in Eq. (3) indicates that the correlation function contains two complex conjugates of the field variable V, namely $V^*(\mathbf{r}_1, t_1)$ and $V^*(\mathbf{r}_2, t_2)$. The second

---

[1] For a general statistical description of stochastic fields in terms of correlation functions of all orders see M&W, Sections 8.2 and 8.3.

superscript indicates that it contains two (non-conjugated) field variables $V(\mathbf{r}_3, t_3)$ and $V(\mathbf{r}_4, t_4)$. In this notation

$$\Gamma^{(1,1)}(\mathbf{r}_1, t_1; \mathbf{r}_2, t_2) = \langle V^*(\mathbf{r}_1, t_1)V(\mathbf{r}_2, t_2)\rangle \tag{4}$$

is the mutual coherence function defined by Eq. (1), when the field ensemble is not necessarily stationary.

In general, no simple relation exists between the correlation functions $\Gamma^{(2,2)}$ and $\Gamma^{(1,1)}$. However, there is an important and broad class of random processes, known as *Gaussian random processes*, for which the correlation functions $\Gamma^{(M,N)}$, with M and N being non-negative integers, can be expressed in terms of the lowest-order one. Such a random process is characterized by the property that the joint probability densities

$$p_n(V_1, V_2, \ldots, V_n; \mathbf{r}_1, t_1; \mathbf{r}_2, t_2; \ldots, \mathbf{r}_n, t_n)$$

are all Gaussian distributions.[1] Random processes of this kind turn up very frequently in nature. A reason for this was mentioned, for example, in connection with Eq. (9) of Section 2.1.

A basic property of a Gaussian random process in any number of variables (known as a *multivariate Gaussian random process*) is that it is completely specified by its first-order and second-order moments (correlations). For example, for such a process, assuming for simplicity that the first moment (the mean) is zero, the following relationship exists:

$$\begin{aligned} \Gamma^{(2,2)}(\mathbf{r}_1, t_1; \mathbf{r}_2, t_2; \mathbf{r}_3, t_3; \mathbf{r}_4, t_4) &\equiv \langle V^*(\mathbf{r}_1, t_1)V^*(\mathbf{r}_2, t_2)V(\mathbf{r}_3, t_3)V(\mathbf{r}_4, t_4)\rangle \\ &= \langle V^*(\mathbf{r}_1, t_1)V(\mathbf{r}_3, t_3)\rangle\langle V^*(\mathbf{r}_2, t_2)V(\mathbf{r}_4, t_4)\rangle \\ &\quad + \langle V^*(\mathbf{r}_1, t_1)V(\mathbf{r}_4, t_4)\rangle\langle V^*(\mathbf{r}_2, t_2)V(\mathbf{r}_3, t_3)\rangle, \end{aligned} \tag{5a}$$

i.e.

$$\begin{aligned} \Gamma^{(2,2)}(\mathbf{r}_1, t_1; \mathbf{r}_2, t_2; \mathbf{r}_3, t_3; \mathbf{r}_4, t_4) &= \Gamma^{(1,1)}(\mathbf{r}_1, t_1; \mathbf{r}_3, t_3)\Gamma^{(1,1)}(\mathbf{r}_2, t_2; \mathbf{r}_4, t_4) \\ &\quad + \Gamma^{(1,1)}(\mathbf{r}_1, t_1; \mathbf{r}_4, t_4)\Gamma^{(1,1)}(\mathbf{r}_2, t_2; \mathbf{r}_3, t_3), \end{aligned} \tag{5b}$$

where $\Gamma^{(1,1)}$ is defined by Eq. (4). As we will see shortly, this is a very important relationship which is of particular relevance to intensity interferometry, which will be discussed in Section 7.3. It is a special case of a general formula known as the *moment theorem*, which holds for any multivariate Gaussian random process. It expresses correlations of any order (M, N) in terms of correlations of order $(1, 1)$.

As we mentioned in Section 2.1, thermal radiation is a realization of a Gaussian random process, which is a consequence of the fact that at room temperature such radiation is largely produced by the process of spontaneous emission.

---

[1] A comprehensive discussion of Gaussian distributions of any number of real or complex variables is given in C.L. Mehta, "Coherence and Statistics of Radiation" in *Lectures in Theoretical Physics*, Vol. VIIC, W.E. Britten ed. (University of Colorado Press, Boulder, CO, 1965), pp. 345–401. See also C.L. Mehta, in *Progress in Optics*, E. Wolf ed. (North-Holland, Amsterdam, 1970), Vol. VIII, Appendix A, pp. 431–434, and K.S. Miller, *Multidimensional Gaussian Distributions* (J. Wiley, New York, 1964).

7.2 Intensity interferometry with radio waves[1]

In the early days of radio astronomy, around 1950, attempts were made to measure the angular diameters of radio stars. It seemed that the use of a Michelson-type stellar interferometer (see Section 3.3.1) for such measurements would not be practical, because it was suspected (incorrectly, as it turned out) that one would have to use extremely long baselines. At that time it was known only that angular diameters of radio stars could not be much greater than a few minutes of arc and some astronomers believed that they might be as small as those of visible stars. At meter wavelengths an interferometer with a baseline of hundreds or perhaps thousands of kilometers would then be required. Obviously, it would not be possible then to use a Michelson-type stellar interferometer, because the required stability could not be maintained over such long distances.

In the early 1950s a British engineer, Robert Hanbury Brown, considered the possibility of using a different type of stellar radio interferometer. In the Michelson stellar interferometer, light arriving from a distant star at two outer mirrors M_1 and M_2 of the interferometer (see Fig. 3.10) propagates to the detection plane, where it forms an interference pattern. From measurements of the fringe visibility, the angular diameter of the star can be deduced by use of the van Cittert–Zernike theorem. In the Hanbury Brown interferometer the signals arriving at the antennas A_1 and A_2 (see Fig. 7.1) are compared with each other by the use of a correlator (multiplier). The first interferometer of this type was described in 1952.[2] It was used to determine the angular diameters of two stars, using antennas separated by a few kilometers. Since that time, the correctness of the results has been confirmed by other interferometers.

The principle of operation of the intensity interferometer with radio waves may be understood from the following considerations. Let

$$I(\mathbf{r}_j, t) = V^*(\mathbf{r}_j, t)V(\mathbf{r}_j, t), \quad (j = 1, 2) \tag{1}$$

be the instantaneous intensities of the radio waves, assumed to be members of a statistically stationary ensemble, arriving at the two antennas A_1 and A_2, located in the neighborhoods of points $\mathbf{r}_1$ and $\mathbf{r}_2$, respectively. For simplicity we will ignore the vector nature of the wavefield, representing it by a scalar wave function. If the mean values $\langle I(\mathbf{r}_1, t)\rangle$ and $\langle I(\mathbf{r}_2, t)\rangle$ of the intensities of the waves arriving at the antennas are subtracted electronically, the intensity fluctuations are given by the expressions

$$\Delta I(\mathbf{r}_j, t) = I(\mathbf{r}_j, t) - \langle I(\mathbf{r}_j, t)\rangle \quad (j = 1, 2). \tag{2}$$

The correlation of the intensity fluctuations at the two antennas is given by the expression

$$\langle \Delta I(\mathbf{r}_1, t)\Delta I(\mathbf{r}_2, t + \tau)\rangle = \langle [I(\mathbf{r}_1, t) - \langle I(\mathbf{r}_1, t)\rangle][I(\mathbf{r}_2, t + \tau) - \langle I(\mathbf{r}_2, t + \tau)\rangle]\rangle, \tag{3}$$

[1] A comprehensive account of intensity interferometry, both with radio waves and with light, is given in R. Hanbury Brown, *The Intensity Interferometer* (Taylor and Francis, London, 1974).
[2] R. Hanbury Brown, R. C. Jennison and M. K. Das Gupta, *Nature* **170** (1952), 1061–1063. See also R. Hanbury Brown and R. Q. Twiss, *Phil. Mag.* **45** (1954), 663–682.

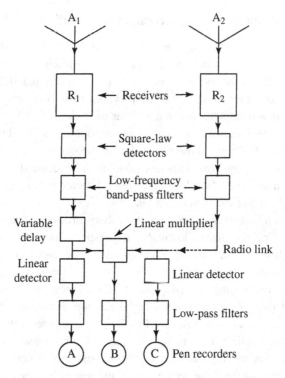

Fig. 7.1 Block diagram of a radio intensity interferometer. [Adapted from R. Hanbury Brown, *The Intensity Interferometer* (Taylor and Francis, London, 1974), p. 84.]

where τ denotes the time delay between the waves arriving at the two antennas. On performing the multiplication on the right-hand side of Eq. (3) and using the fact that the assumed stationarity of the field ensemble implies that $\langle I(\mathbf{r}_2, t + \tau)\rangle = \langle I(\mathbf{r}_2, t)\rangle$, Eq. (3) gives

$$
\begin{aligned}
\langle \Delta I(\mathbf{r}_1, t)\Delta I(\mathbf{r}_2, t + \tau)\rangle &= \langle I(\mathbf{r}_1, t)I(\mathbf{r}_2, t + \tau)\rangle - \langle I(\mathbf{r}_1, t)\rangle\langle I(\mathbf{r}_2, t)\rangle \\
&\quad - \langle I(\mathbf{r}_1, t)\rangle\langle I(\mathbf{r}_2, t)\rangle + \langle I(\mathbf{r}_1, t)\rangle\langle I(\mathbf{r}_2, t)\rangle \\
&= \langle I(\mathbf{r}_1, t)I(\mathbf{r}_2, t + \tau)\rangle - \langle I(\mathbf{r}_1, t)\rangle\langle I(\mathbf{r}_2, t)\rangle. \quad (4)
\end{aligned}
$$

Let us express the right-hand side of Eq. (4) in terms of the fluctuating field incident on the antennas. One then finds that

$$
\begin{aligned}
\langle \Delta I(\mathbf{r}_1, t)\Delta I(\mathbf{r}_2, t + \tau)\rangle &= \langle V^*(\mathbf{r}_1, t)V(\mathbf{r}_1, t)V^*(\mathbf{r}_2, t + \tau)V(\mathbf{r}_2, t + \tau)\rangle \\
&\quad - \langle V^*(\mathbf{r}_1, t)V(\mathbf{r}_1, t)\rangle\langle V^*(\mathbf{r}_2, t)V(\mathbf{r}_2, t)\rangle. \quad (5)
\end{aligned}
$$

We note that the first term on the right is the fourth-order correlation function which we encountered earlier [Eq. (3) of Section 7.1]:

$$
\langle V^*(\mathbf{r}_1, t)V(\mathbf{r}_1, t)V^*(\mathbf{r}_2, t + \tau)V(\mathbf{r}_2, t + \tau)\rangle \equiv \Gamma^{(2,2)}(\mathbf{r}_1, t, \mathbf{r}_2, t + \tau; \mathbf{r}_1, t, \mathbf{r}_2, t + \tau). \quad (6)
$$

It is reasonable to assume that the wavefield reaching the antennas arises from the super-position of many independent contributions of the stellar radio source; and consequently, according to the central limit theorem, mentioned in connection with Eq. (9) of Section 2.1, the field arriving at the antennas will be a realization of a Gaussian random process. Under these circumstances, and assuming the wavefield to be statistically stationary, the fourth-order correlation function (6) may be expressed, according to Eq. (5a) of Section 7.1, in the form

$$
\begin{aligned}
\langle V^*(\mathbf{r}_1, t)&V(\mathbf{r}_1, t)V^*(\mathbf{r}_2, t + \tau)V(\mathbf{r}_2, t + \tau)\rangle \\
&= \langle V^*(\mathbf{r}_1, t)V(\mathbf{r}_1, t)\rangle\langle V^*(\mathbf{r}_2, t + \tau)V(\mathbf{r}_2, t + \tau)\rangle \\
&\quad + \langle V^*(\mathbf{r}_1, t)V(\mathbf{r}_2, t + \tau)\rangle\langle V^*(\mathbf{r}_2, t + \tau)V(\mathbf{r}_1, t)\rangle \\
&= \langle I(\mathbf{r}_1, t)\rangle\langle I(\mathbf{r}_2, t)\rangle + \Gamma^{(1,1)}(\mathbf{r}_1, \mathbf{r}_2, \tau)\Gamma^{(1,1)}(\mathbf{r}_2, \mathbf{r}_1, -\tau)
\end{aligned}
\tag{7}
$$

or, since $\Gamma^{(1,1)}$ obeys the relation

$$
\Gamma^{(1,1)}(\mathbf{r}_2, \mathbf{r}_1, -\tau) = \Gamma^{(1,1)*}(\mathbf{r}_1, \mathbf{r}_2, \tau),
\tag{8}
$$

one has

$$
\begin{aligned}
\langle V^*(\mathbf{r}_1, t)&V(\mathbf{r}_1, t)V^*(\mathbf{r}_2, t + \tau)V(\mathbf{r}_2, t + \tau)\rangle \\
&= \langle I(\mathbf{r}_1, t)\rangle\langle I(\mathbf{r}_2, t)\rangle + |\Gamma^{(1,1)}(\mathbf{r}_1, \mathbf{r}_2, \tau)|^2.
\end{aligned}
\tag{9}
$$

On substituting from Eq. (9) into Eq. (5) we obtain for the correlation of the intensity fluctuations the formula

$$
\langle \Delta I(\mathbf{r}_1, t)\Delta I(\mathbf{r}_2, t + \tau)\rangle = |\Gamma^{(1,1)}(\mathbf{r}_1, \mathbf{r}_2, \tau)|^2.
\tag{10}
$$

Now the second-order correlation function $\Gamma^{(1,1)}(\mathbf{r}_1, \mathbf{r}_2, \tau)$ which appears on the right-hand side of Eq. (10) is just the mutual coherence function which we denoted before by $\Gamma(\mathbf{r}_1, \mathbf{r}_2, \tau)$; its normalized version

$$
\frac{\Gamma(\mathbf{r}_1, \mathbf{r}_2, \tau)}{\sqrt{\Gamma(\mathbf{r}_1, \mathbf{r}_1, 0)}\sqrt{\Gamma(\mathbf{r}_2, \mathbf{r}_2, 0)}} = \frac{\Gamma(\mathbf{r}_1, \mathbf{r}_2, \tau)}{\sqrt{\langle I(\mathbf{r}_1, t)\rangle\langle I(\mathbf{r}_2, t)\rangle}}
$$

$$
= \gamma(\mathbf{r}_1, \mathbf{r}_2, \tau)
\tag{11}
$$

is just the degree of coherence of the wavefield at the points $\mathbf{r}_1$ and $\mathbf{r}_2$. Hence Eq. (10) may be expressed in the form

$$
\langle \Delta I(\mathbf{r}_1, t)\Delta I(\mathbf{r}_2, t + \tau)\rangle = \langle I(\mathbf{r}_1, t)\rangle\langle I(\mathbf{r}_2, t)\rangle|\gamma(\mathbf{r}_1, \mathbf{r}_2, \tau)|^2,
\tag{12}
$$

which implies that

$$
\frac{\langle \Delta I(\mathbf{r}_1, t)\Delta I(\mathbf{r}_2, t + \tau)\rangle}{\langle I(\mathbf{r}_1, t)\rangle\langle I(\mathbf{r}_2, t)\rangle} = |\gamma(\mathbf{r}_1, \mathbf{r}_2, \tau)|^2.
\tag{13}
$$

Equation (13) is the basic formula of intensity interferometry for thermal radiation. It shows that the absolute value of the degree of coherence of a thermal field at two field points $\mathbf{r}_1$ and $\mathbf{r}_2$ may be determined from measurements of the correlations of intensity fluctuations and the average intensities at these points.

We recall that by use of the Michelson stellar interferometer one can, in principle, determine the complex degree of coherence $\gamma(\mathbf{r}_1, \mathbf{r}_2, \tau)$ at the two mirrors from visibility measurements although, in practice, atmospheric turbulence essentially destroys information about its phase. In intensity interferometry, as we just showed, one can determine the modulus $|\gamma(\mathbf{r}_1, \mathbf{r}_2, \tau)|$ of the degree of coherence. By use of the van Cittert–Zernike theorem one can then obtain information about the intensity distribution across the source, assuming that it is spatially incoherent. However, as is evident from Eq. (13), such measurements cannot provide, even in principle, any information about the phase of the degree of coherence. This limitation is, however, of no consequence, if the stellar disk is rotationally symmetric, as we briefly discussed in Section 3.3.1.

7.3 The Hanbury Brown–Twiss effect and intensity interferometry with light

After the first successful determination of the angular diameters of radio stars by intensity interferometry, Hanbury Brown and Twiss turned their attention to the possibility of using a similar technique to measure the angular diameters of stars at optical wavelengths. It was clear that, if this were possible, an appreciable amount of valuable new information could be obtained. Up to that time the only way to measure stellar diameters at optical wavelengths was by use of the Michelson stellar interferometer. Practical difficulties had restricted the use of that method to measurements made by Michelson and Pease in the 1920s determining the angular diameters of only six stars. A source of the difficulties in such measurements was the extremely small angles subtended by stars at the Earth's surface – of the order of 10^{-4} seconds of arc; this requires the use of interferometers with baselines of several hundred meters. Further, atmospheric turbulence produces blurring and distortions of the image of the star. It was chiefly for these reasons that no progress made in applying Michelson's method to determining the angular diameters of other stars than the small number initially measured.

At first glance it seemed that intensity interferometry would overcome some of the limitations of Michelson's method and this indeed turned out to be the case. However, the theory of intensity interferometry with light rather than with radio waves turned out to be rather subtle, mainly due to the quantum-mechanical nature of detection of light fluctuations. To begin with we will ignore this complication and we will assume that Eq. (13) of Section 7.2, which we derived for intensity interferometry with radio waves, applies even when a very different detection process is employed, using the photoelectric effect. In Section 7.5 we will explain why Eq. (10) of Section 7.2, which we derived on the implicit assumption that the detector measures correlations in the intensity fluctuations of the classical field, can be used even when photoelectric detectors are used, as long as some subtleties connected with the signal-to-noise ratio are ignored. Such subtleties must, of course, be taken into account in practice, but omitting this aspect of the technique does not prevent an understanding of the basic physical principles underlying it.

In the spectral region of radio waves, the procedure of determining the correlation in the intensity fluctuations is straightforward, because square-law detectors measuring intensity fluctuations $\Delta I(\mathbf{r}, t)$ of radio waves are well-known electronic devices. One need only set up two radio antennas in the neighborhood of the points $\mathbf{r}_1$ and $\mathbf{r}_2$. The output signals are then multiplied in an electronic correlator and the product is averaged and recorded.

The situation is quite different with light. Light detectors with a sufficiently rapid response are photoelectric detectors. Such detectors are, however, highly "non-classical" devices, making use of the photoelectric effect. This is the phenomenon of ejection of electrons from a metal when electromagnetic radiation of short enough wavelength impinges on its surface. It has been known for a long time that the energy of the ejected electron is independent of the intensity of the light that illuminates the metal surface but depends on the frequency of the light. When the intensity is increased, the number of electrons ejected increases, but not their energy. These observations are in contradiction with the classical wave theory of light. Einstein, in a famous paper published in 1905, explained these observations on the basis of the corpuscular nature of light, i.e. by postulating that light incident on the photoelectric surface consists of particles – light quanta – now called photons. Einstein's paper was the first paper which clearly indicated the need for quantizing the electromagnetic field under some circumstances. In 1921 he was awarded the Nobel Prize for Physics "for his services to Theoretical Physics and especially for his discovery of the law of the photoelectric effect."

We derived the basic formula (13) of Section 7.2 using classical wave theory for the detection of radio waves. However, because of the non-classical nature of the photoelectric effect, it is by no means clear that the formula would also hold for light waves detected by use of this effect. In fact there was a bitter controversy, which we will mention later, regarding this question; it was eventually resolved in favor of the validity of the classical formula even under these circumstances. The justification involved some beautiful physics relating to the wave–particle duality of light which we will discuss in Section 7.4; it also required the development of the theory of photoelectric detection of light fluctuations, which was formulated largely in order to clarify some of the questions surrounding this problem.

In 1956, shortly after the successful determination of diameters of radio stars by intensity interferometry, Hanbury Brown and Twiss performed laboratory experiments to determine whether the technique worked also with light.[1] Light from a mercury arc was filtered and split into two beams by a half-silvered mirror and the two beams were then incident on two photo-multiplier tubes, one of which was movable across the beam (see Fig. 7.2). As the photo-multiplier moved, the degree of coherence of the light at the two photo-cathodes varied in a manner that could be calculated from the geometry of the arrangement. The electrical signals from the two photo-multipliers were fed into electronic circuits, where they

[1] R. Hanbury Brown and R. Q. Twiss, *Nature* **177** (1956), 27–29.

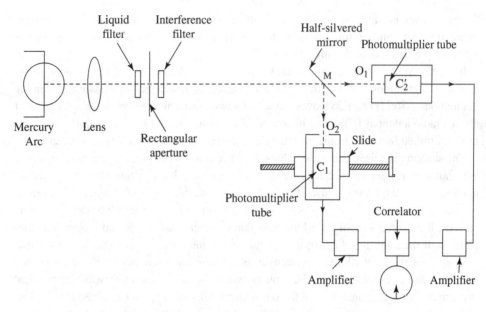

Fig. 7.2 Apparatus used for the demonstration of the Hanbury Brown–Twiss effect. [Adapted from R. Hanbury Brown and R. Q. Twiss, *Nature* **177** (1956), 27–29.]

were amplified. The outputs were multiplied together in a multiplier. With this equipment the intensity correlation was measured for various positions of the movable photo-multiplier and compared with the theoretical values calculated from the geometry of the system, the spectral distribution of the light and the electrical characteristics of the circuits. The results are shown in Fig. 7.3 and are seen to agree well with predictions of the classical theory.[1]

In spite of the successful laboratory experiment a controversy surrounding the effect continued and papers were published reporting results of similar experiments but with negative results, suggesting that the effect does not exist. It was shown later that the negative results were largely due to substantial underestimation of the time needed to obtain a significant signal-to-noise ratio.[2]

Before discussing the resolution of the controversy surrounding the Hanbury Brown–Twiss effect and outlining why the "classical theory" applies to understanding the principles of the optical intensity interferometer, we will briefly describe the first two such interferometers, which were built in the period 1956–1966.

[1] A similar experiment was performed by W. Martienssen and E. Spiller, *Am. J. Phys.* **32** (1964), 919–926 by using so-called pseudo-thermal sources. Such sources were produced from laser sources by changing the statistical properties of the light by passing it through a rotating diffuser. In this way light was generated whose statistical properties were governed by Gaussian distributions; but, unlike the usual thermal sources, it had a very high statistical degeneracy (see Appendix I), which made it appreciably easier to demonstrate the effect.

[2] It was estimated that in one of the experiments [Á. Adam, L. Jánossy and P. Varga, *Ann. Phys.* (Leipzig) **16** (1955), 408–413] 10^{11} years (longer than the estimated age of the Earth) would have been needed to obtain a significant signal-to-noise ratio. In experiments described by E. Brannen and H. I. S. Ferguson, *Nature* **178** (1956), 481–482, the estimated time required for obtaining a significant signal-to-noise ratio was 1,000 years.

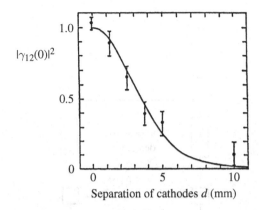

Fig. 7.3 The experimental and theoretical values of the squared modulus of the degree of coherence $|\gamma_{12}(0)|^2$ for various separations between the photo-cathodes. [Adapted from R. Hanbury Brown and R. Q. Twiss, *Proc. Roy. Soc.* (London) **A243** (1958), 291–319.]

A block diagram of the detector part of an optical stellar-intensity interferometer is shown in Fig. 7.4. The first interferometer of this kind[1] utilized devices that were somewhat primitive, namely two parabolic reflectors of searchlights used by the British Army. Their chief purpose was to concentrate light from a star onto two photodetectors, without aiming at producing a good image. For, as we have already noted, the phase of the incident light plays no role in the determination of the angular diameter of the star by this technique, because the correlation of the intensity fluctuations at the two detectors depend only on the modulus of the degree of coherence of the incident light, not on its phase [see Eq. (13) of Section 7.2]. For the same reason the atmospheric fluctuations are of less consequence in such measurements than in measurements that use the Michelson stellar interferometer.

Using this interferometer Hanbury Brown and Twiss found that the angular diameter of the star Sirius is 6.9×10^{-3} seconds of arc, using baselines of up to about 9 m (see Fig. 7.5).

Having convincingly demonstrated the possibility of determining the angular diameters of visual stars by intensity interferometry, Hanbury Brown and Twiss and their collaborators built a large interferometer of this kind at Narrabri in Australia, about 370 miles by road north of Sydney. The general layout of the interferometer is shown in Fig. 7.6 and can best be described by quoting the following passage from Hanbury Brown's book (p. 94) on this subject cited earlier:

> The photoelectric detectors were each mounted at the focus of two very large reflectors carried on trucks running on a circular railway track with a gauge of 5.5 m and a diameter of 188 m. These mobile trucks were connected to the control building by cables suspended from steel catenary wires which were attached at one

[1] R. Hanbury Brown and R. Q. Twiss, *Nature* **178** (1956), 1046–1048; *Proc. Roy. Soc.* (London), **A248** (1958), 222–237.

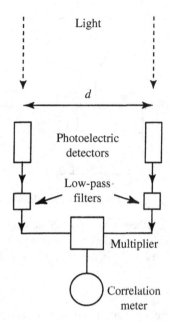

Light

d

Photoelectric
detectors

Low-pass
filters

Multiplier

Correlation
meter

Fig. 7.4 Block diagram of an optical stellar intensity interferometer. [Adapted from R. Hanbury Brown, *The Intensity Interferometer* (Taylor and Francis, London, 1974), p. 48.]

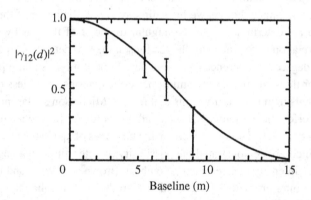

Fig. 7.5 The squared modulus of the degree of coherence of light from the star Sirius A. The points indicate measured values, together with probable errors. The curve shows the theoretical values for a star of angular diameter 0.0069 seconds of arc. [Adapted from R. Hanbury Brown and R. Q. Twiss, *Proc. Roy. Soc.* (London) **A248** (1958), 222–237.]

end to a bearing at the top of a tower in the center of the circle, and at the other to a small tender towed by each truck. When not in use the reflectors were housed in a garage built over the southern sector. A valuable but expensive feature of this garage was a slot running almost the full length of one wall enabling the trucks to

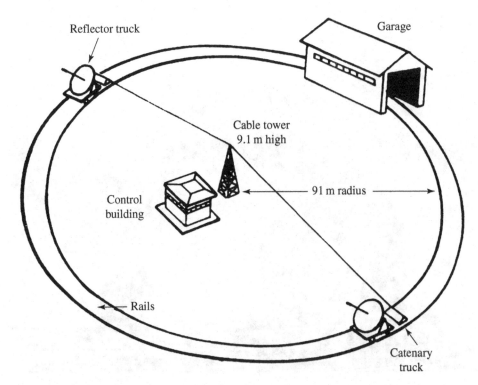

Fig. 7.6 The general layout of the stellar intensity interferometer at Narrabri. [Adapted from R. Hanbury Brown, *Sky & Telescope* **28** (1964), 64–69.]

be parked inside without detaching the cables and hence without disturbing the electric connections.

The control building which housed the control desk, the computer, a large air-conditioning plant and various motor generators, switchboards, etc. was a two-story brick building of solid construction with a good heat-reflecting roof so that it was possible even in the summer at Narrabri to hold the inside temperature to $72 \pm 2°F$. It was close enough to the central tower to allow the catenary cables to pass over its roof [see Fig. 7.6].

The reflectors [shown in Fig. 7.7] were regular 12-sided polygons roughly 6.5 m in diameter, each having a useable reflecting area of 30 m^2. They were mounted on turntables carried by trucks.

The focal length of each mirror was about 11 m and the surface of each reflector was a mosaic of 252 hexagonal mirrors (Fig. 7.7), each approximately 38 cm between opposite sides and 2 cm thick. The angular diameters of 32 stars were determined by this instrument and this information represents a major contribution to astronomy.

Although the Hanbury Brown–Twiss effect was discovered, and has been used, in connection with determining stellar diameters, it has found other applications since then, for

Fig. 7.7 Twin light collectors of the stellar-intensity interferometer on the circular track at Narrabri Observatory, New South Wales, Australia. Some of the small mirrors have been removed from each of the 6.5-m reflectors for recoating. A beam 11 m long supports a photoelectric detector at the focus of each reflector. [Adapted from R. Hanbury Brown, *Sky & Telescope* **28** (1964), 64–69.]

example in high-energy physics, nuclear physics, atomic physics and condensed-matter physics.[1]

7.4 Einstein's formula for energy fluctuations in blackbody radiation and the wave–particle duality

Let us now return to the question of why the performance of an optical intensity interferometer can be described by classical wave theory, even though the photoelectric detectors act in a highly non-classical manner. A turning point in the controversy surrounding this subject was the publication of a short note by E. M. Purcell.[2] He showed by a simple heuristic

---

[1] See, for example, G. Baym, *Acta Phys. Polon. B* **29** (1998), 1839–1884; M. Schellekens, R. Hoppeler, A. Perrin, J. Viana Gomez, D. Boiron, A. Aspect and C. I. Westbrook, *Science* **310** (2005), 648–651; A. Öttl, S. Ritter, M. Köhl and T. Esslinger, *Phys. Rev. Lett.* **95** (2005), 090404.
[2] E. M. Purcell, *Nature* **178** (1956), 1449–1450.

argument that, when a beam of light is incident on a photodetector, the statistical fluctuations in the counting rate of the emitted electrons will be greater than one would expect in a random sequence of independent events occurring at the same average rate. The additional fluctuations give rise to the cross-correlation in the intensity fluctuations found by Hanbury Brown and Twiss, which is essentially a classical wave contribution.

The presence of both a particle contribution and a wave contribution in the photoelectric detection of light fluctuations brought out by Purcell's analysis may be regarded as a manifestation of the wave–particle duality, which is a basic feature of quantum mechanics. It was discovered by Einstein[1] well before the formulation of quantum mechanics, in a beautiful paper dealing with energy fluctuations in blackbody radiation. Because Einstein's paper has a close bearing on Purcell's argument regarding the origin of the Hanbury Brown–Twiss effect, we will briefly present its essence here.

Einstein considered blackbody radiation in equilibrium in a cavity of large volume V. In any subregion of volume $v \ll V$ the energy will fluctuate, moving in and out of the subregion. He inquired about the magnitude of the fluctuations in this region in a small frequency range $(v, v + dv)$, choosing as the measure of the fluctuations the variance of the energy E in that region, i.e. the quantity

$$\overline{(\Delta E)^2} = \overline{(E - \bar{E})^2}, \tag{1}$$

where, for convenience, we now use an overbar rather than angular brackets for an ensemble average. Einstein showed first from thermodynamic considerations that[2]

$$\overline{(\Delta E)^2} = k_B T^2 \frac{\partial \bar{E}}{\partial T}, \tag{2}$$

where k_B is the Boltzmann constant and T is the absolute temperature. Equation (2) is often called the *Einstein–Fowler formula*.

Now, for blackbody radiation, $\bar{E}$ is given by Planck's law

$$\bar{E} = Z \frac{hv}{e^{hv/(k_B T)} - 1}, \tag{3}$$

where h is Planck's constant and

$$Z = \frac{8\pi V v^2 \, dv}{c^3}, \tag{4}$$

c being the speed of light in vacuum. The quantity Z may be shown to represent the so-called number of cells in phase space associated with the volume V in ordinary space and

---

[1] A. Einstein, *Phys. Z.* **10** (1909), 185–193; English translation in *The Collected Papers of Albert Einstein*, Vol. 2 (Princeton University Press, Princeton, NJ, 1989), 357–375.

[2] For a derivation of this formula see, for example, F. Reiche, *The Quantum Theory* (Methuen, London, 1922), Note 52, 139–140.

the frequency range $(\nu, \nu + d\nu)$ [see Appendix I(b)]. On substituting for the mean energy $\bar{E}$ from Eq. (3) into Eq. (2), one obtains for the variance of the energy fluctuations in the subregion the expression

$$\overline{(\Delta E)^2} = h\nu\bar{E} + \frac{1}{Z}\bar{E}^2. \tag{5}$$

This formula is known as *Einstein's fluctuation formula for blackbody radiation*.

Einstein drew some remarkable conclusions from this formula. He pointed out that if the cavity V contains only waves, as most physicists at that time believed, one could decompose the field within the cavity into a set of plane-wave modes of different amplitudes and phases, propagating in different directions. Short-term interference of the waves would then produce field fluctuations and, consequently, energy fluctuations in the subregion v of the cavity. Einstein argued, by using a simple dimensional argument, that the variance of these fluctuations will be given exactly by the second term in Eq. (5), namely by the formula[1]

$$\overline{(\Delta E)^2_{\text{waves}}} = \frac{1}{Z}\bar{E}^2. \tag{6}$$

Einstein next argued as follows: since only the second term on the right-hand side of Eq. (5) can be understood on the basis of wave theory, that theory does not fully explain the origin of the fluctuations. Suppose that the volume contains n particles (quanta), each of energy $h\nu$. If they were independent classical particles, the fluctuations in their number would obey the Poisson distribution [see Appendix IV(a)], viz.,

$$p(n) = \frac{\bar{n}^n e^{-\bar{n}}}{n!}. \tag{7}$$

---

[1] Einstein's heuristic argument by means of which he interpreted the meaning of the second term on the right of his fluctuation formula (5) occupies only eight lines of his paper. Another plausibility argument, indicating that the variance of fluctuations of the waves is proportional to $\bar{E}^2$, may readily be given on the basis of Eq. (12) of Section 7.2. That formula is an expression for the correlation of intensity fluctuations in a field governed by a multivariate Gaussian distribution (i.e. a Gaussian random process), as might be expected to be the case for blackbody radiation. If we set in that formula $\mathbf{r}_1 = \mathbf{r}_2 = \mathbf{r}$ and $\tau = 0$, then $\gamma(\mathbf{r}, \mathbf{r}, 0) = 1$ and the equation reduces to (using now an overbar instead of angular brackets to denote an ensemble average)

$$\overline{(\Delta I)^2} = \bar{I}^2,$$

where I is the instantaneous intensity. Evidently I may be taken to be a measure of the instantaneous energy E. Writing $E = \alpha I$, where α is a constant, the above formula reduces to

$$\overline{(\Delta E)^2} \sim \bar{E}^2.$$

A long but rigorous derivation of the formula (6) was given by H. A. Lorentz in his book *Les théories statistiques en thermodynamique* (Teubner, Leipzig and Berlin, 1916), 114–120.

The total energy of n such particles would be $E = nh\nu$, so that

$$n = \frac{E}{h\nu}. \tag{8}$$

Now the variance of a Poisson distribution is

$$\overline{(\Delta n)^2} = \bar{n}, \tag{9}$$

and it follows from Eqs. (8) and (9) that the variance $\overline{[\Delta(E/(h\nu))]^2} = \bar{E}/(h\nu)$, i.e.

$$\overline{(\Delta E)^2_{\text{particles}}} = h\nu\bar{E}. \tag{10}$$

Here we have attached the subscript "particles" to stress that the expression was derived on the assumption that the energy fluctuations are caused by particles. We see that the expression on the right of Eq. (10) is precisely the first term on the right-hand side of Eq. (5). Hence, using also Eq. (6), we can express Eq. (5) as

$$\overline{(\Delta E)^2} = \overline{(\Delta E)^2}_{\text{particles}} + \overline{(\Delta E)^2}_{\text{waves}}. \tag{11}$$

This result was the first example of the wave–particle duality, which about 15 years later became a prominent feature of quantum mechanics.

Equation (11) is a kind of a hybrid result that expresses the variance of energy fluctuations in a cavity containing blackbody radiation as a sum of contributions of classical particles and classical waves. Today one would interpret Einstein's result somewhat differently, in terms of either non-classical waves or non-classical particles. For example, by using Eq. (8) we could rewrite Eq. (5) in the form

$$\overline{(\Delta n)^2} = \bar{n} + \frac{1}{Z}\bar{n}^2. \tag{12}$$

The presence of the second term on the right of this formula shows that the particles do not behave as classical particles which obey a Poisson distribution because, for such particles, Eq. (9) rather than (12) would apply. Equation (12) for the variance is, in fact, a well-known formula pertaining to quantum particles known as bosons in Z cells of phase space [see Appendix I(b)].

With this background we will now return to the Hanbury Brown–Twiss experiment and consider, more generally, the theory of photoelectric detection of light fluctuations.

7.5 Mandel's theory of photoelectric detection of light fluctuations

7.5.1 Mandel's formula for photocount statistics

As noted in the previous section, the controversy surrounding the theory of photoelectric detection of light fluctuations was largely resolved by a brief discussion by E. M. Purcell.

He showed that the reason why correlations in the fluctuations of a classical optical field can be extracted from photocount measurements was due to the excess fluctuations (arising from the boson nature of light) over completely random counts.

L. Mandel in two important papers[1] carried Purcell's argument further and developed a theory of photoelectric detection of light fluctuations, which he appropriately referred to as *stochastic association of photons with random waves*. Mandel's analysis not only gave a clear explanation of the Hanbury Brown–Twiss effect, but also provided a general theory of the interaction of stochastic light beams with photoelectric detectors. We will now give a brief account of it.

Consider first a single photodetector exposed to an incident light wave, assumed for simplicity to be linearly polarized, with fluctuating complex amplitude $V(t)$, represented by an analytic signal (Section 2.3). It seems plausible and can be shown, for example, by semiclassical theory,[2] that the probability that an electron will be emitted within a time interval $(t, t + \Delta t)$ is proportional to the instantaneous light intensity $I(t) = V^*(t)V(t)$, i.e.

$$P(t)\Delta t = \alpha I(t)\Delta t, \tag{1}$$

α being a constant representing the quantum efficiency of the photodetector. Starting from Eq. (1) Mandel showed that the probability that n electrons are emitted within a time interval $(t, t + T)$ is given by the formula

$$p(n, t, T) = \frac{1}{n!}[\alpha W(t, T)]^n e^{-\alpha W(t,T)}, \tag{2}$$

where

$$W(t, T) = \int_t^{t+T} I(t')dt' \tag{3}$$

is the integrated intensity of the light incident on the detector over the time interval $(t, t + T)$. The derivation of Eq. (2) is somewhat lengthy. It is given in Appendix III.

It is of importance to appreciate that Eq. (2) pertains to a single realization of an incident, generally random, field. Physically more significant is the average, which we will denote by an overbar, of the expression given in Eq. (2) over the ensemble of the incident field:

$$P(n, t, T) = \overline{p(n, t, T)}$$
$$= \frac{1}{n!}\overline{[\alpha W(t, T)]^n e^{-\alpha W(t,T)}} \tag{4}$$

or, more explicitly,

$$P(n, t, T) = \int_0^\infty \frac{[\alpha W(t, T)]^n}{n!} e^{-\alpha W(t,T)} p(W)dW, \tag{5}$$

[1] L. Mandel, *Proc. Phys. Soc.* (London) **72** (1958), 1037–1048; and *Ibid.* **74** (1959), 223–243.
[2] L. Mandel, E. C. G. Sudarshan and E. Wolf, *Proc. Phys. Soc.* (London) **84** (1964), 435–444, Eq. (2.18b).

where $p(W) \equiv p(W, t, T)$ is the probability density of the integrated intensity, given by Eq. (3), the variables t and T being kept fixed in the integration. We will refer to Eq. (5) as *Mandel's formula for photocount statistics*.

The formula is of considerable generality. It applies to any kind of linearly polarized incident light, irrespective of its statistical properties and irrespective of whether the light fluctuations are statistically stationary or not. It should be noted that the formula involves probabilities in two different ways. First, it contains the probability relating to emission of photoelectrons from a single realization of the incident field; and second, it involves the ensemble of the incident field.

The integrand on the right-hand side of Mandel's formula contains the expression $[\alpha W(t, T)]^n e^{-\alpha W(t, T)}/n!$, which is a Poisson distribution with mean

$$\bar{n} = \alpha W(t, T). \tag{6}$$

However, Mandel's formula (5) is, in general, not a Poisson distribution, because of the presence of the weighting factor $p(W)$ under the integral sign. The formula may be said to represent a *Poisson transform* of the probability $p(W)$.

7.5.2 *The variance of counts from a single photodetector*

The variance of the counts, recorded in the time interval $(t, t + T)$ is given by the formula

$$\overline{(\Delta n)^2} = \overline{(n - \bar{n})^2} = \overline{n^2} - \bar{n}^2, \tag{7}$$

where the first two moments,

$$\bar{n} = \sum_{n=0}^{\infty} n\, \overline{p(n, t, T)} \tag{8a}$$

and

$$\overline{n^2} = \sum_{n=0}^{\infty} n^2\, \overline{p(n, t, T)}, \tag{8b}$$

of the averaged Poisson distribution $\overline{p(n, t, T)}$, defined by Eq. (4), are given by the well-known formulas[1]

$$\bar{n} = \alpha \overline{W(t, T)}, \tag{9a}$$

$$\overline{n^2} = \alpha \overline{W(t, T)} + \alpha^2 \overline{[W(t, T)]^2}. \tag{9b}$$

---

[1] See, for example, A. Papoulis, *Probability, Random Variables and Stochastic Processes* (McGraw-Hill, New York, 1965), 145.

Hence the variance defined by Eq. (7) is given by the expression

$$\overline{(\Delta n)^2} = \alpha \overline{W(t,T)} + \alpha^2 \overline{[W(t,T)]^2} - \alpha^2 \left| \overline{W(t,T)} \right|^2, \tag{10}$$

or, using Eq. (9a),

$$\overline{(\Delta n)^2} = \bar{n} + \alpha^2 \overline{[\Delta W(t,T)]^2}, \tag{11}$$

where

$$\overline{[\Delta W(t,T)]^2} = \overline{\left| W(t,T) - \overline{W(t,T)} \right|^2} = \overline{[W(t,T)]^2} - \left[\overline{W(t,T)} \right]^2 \tag{12}$$

is the variance of the integrated intensity $W(t,T)$ defined by Eq. (3).

Equation (11) has a simple interpretation. It shows that the variance of the fluctuations in the number of ejected photoelectrons may be regarded as having two separate contributions: (i) from the fluctuations in the number of particles obeying the classical Poisson distribution (the term $\bar{n}$) and (ii) from fluctuations of a classical wave (the wave interference term $\alpha^2 \overline{[\Delta W(t,T)]^2}$). This result which, as we have just shown, holds for any radiation field, is strictly analogous to the celebrated result of Einstein, relating to energy fluctuations in a region of a cavity containing blackbody radiation, under conditions of thermal equilibrium, as discussed in Section 7.4. We now see that a fluctuation formula of the kind Einstein derived, which clearly exhibits the wave–particle duality of radiation, also holds for counting fluctuations in *time intervals* for any light beam (i.e. thermal or non-thermal, stationary or non-stationary), at points that may be situated far away from the source of the light.

It is of interest to note that when the intensity of the incident beam is stabilized, e.g. as in a single-mode laser beam, the variance $[\Delta W(t,T)]^2$ will be negligible because there will be no intensity fluctuations in the incoming beam and the formula (11) then reduces to $\overline{(\Delta n)^2} = \bar{n}$], as for a system of classical particles.

If, on the other hand, the light is of thermal origin, the probability density $p(n,t,T)$ of the photocount will be quite different. For thermal light the probability distribution of the incident field will be Gaussian (see Eq. (9) of Section 2.1). By using elementary rules of the theory of probability, this may be shown to imply that the probability distribution of the intensity is exponential (see M&W, Section 3.1.4), i.e. that

$$p(I) = \frac{1}{\bar{I}} \exp(-I/\bar{I}). \tag{13}$$

One may then show by the use of Mandel's formula (5) and formulas (9a) and (3) that if the integration time T is much shorter than the coherence time of the light, i.e. if $T \ll 2\pi/\Delta\omega$, where $\Delta\omega$ is the bandwidth of the incident light, the probability distribution of the photo-electron becomes[1]

$$P(n, t, T) = \overline{p(n, t, T)} = \frac{\bar{n}^n}{(\bar{n} + 1)^{n+1}}, \tag{14}$$

which is the Bose–Einstein distribution for one cell of phase space (see Appendix IV, Eq. (16), or Morse's discussion[2]).

7.5.3 Correlation between count fluctuations from two detectors

We will now consider the situation when a light beam illuminates two photodetectors, a situation which, as we saw, is of special interest in optical intensity interferometry. Suppose that the detectors are located in the neighborhoods of points $\mathbf{r}_1$ and $\mathbf{r}_2$, respectively. We will indicate by subscripts 1 and 2 the outputs of each of the two detectors.

Let us determine the correlation $\overline{\Delta n_1 \Delta n_2}$ of the fluctuations

$$\Delta n_1 = n_1 - \bar{n}_1 \quad \text{and} \quad \Delta n_2 = n_2 - \bar{n}_2 \tag{15}$$

of the photocounts of the two detectors. We have

$$\begin{aligned}
\overline{n_1 n_2} &= \sum_{n_1=0}^{\infty} \sum_{n_2=0}^{\infty} \overline{n_1 n_2 P_1(n_1, t_1, T) p_2(n_2, t_2, T)} \\
&= \sum_{n_1=0}^{\infty} \overline{n_1 p_1(n_1, t_1, T)} \sum_{n_2=0}^{\infty} \overline{n_2 p_2(n_2, t_2, T)},
\end{aligned} \tag{16}$$

where the probabilities $p_j(n_j, t_j, T)$, $j = 1, 2$, are given by expressions such as Eq. (2). Now, according to Eqs. (8a) and (9a),

$$\bar{n}_1 \equiv \sum_{n_1=0}^{\infty} \overline{n_1 P_1(n_1, t_1, T)} = \alpha_1 \overline{W_1(t_1, T)}, \tag{17a}$$

$$\bar{n}_2 \equiv \sum_{n_2=0}^{\infty} \overline{n_2 p_2(n_2, t_2, T)} = \alpha_2 \overline{W_2(t_2, T)}, \tag{17b}$$

i.e. the expectation values of n_1 and n_2 are proportional to the averaged integrated intensity $\overline{W(t, T)}$ at each of the two points. Hence, using Eqs. (17), we have

$$\bar{n}_1 \bar{n}_2 = \alpha_1 \alpha_2 \overline{W_1(t_1, T)} \, \overline{W_2(t_2, T)}. \tag{18}$$

[1] L. Mandel in *Progress in Optics*, E. Wolf ed. (North-Holland, Amsterdam, 1963), Vol. 2, p. 228, Eq. (71b).
[2] P. M. Morse, *Thermal Physics* (Benjamin, New York, 1962), p. 218.

Now

$$\overline{\Delta n_1 \, \Delta n_2} = \overline{(n_1 - \bar{n}_1)(n_2 - \bar{n}_1)}$$

$$= \overline{n_1 n_2} - \bar{n}_1 \bar{n}_2 \tag{19}$$

or, using Eqs. (16) and (18),

$$\overline{\Delta n_1 \, \Delta n_2} \equiv \alpha_1 \alpha_2 \overline{[W_1(t_1, T)W_2(t_2, T)]} - \alpha_1 \alpha_2 \overline{W_1(t_1, T)} \; \overline{W_2(t_2, T)},$$

which implies that

$$\overline{\Delta n_1 \, \Delta n_2} = \alpha_1 \alpha_2 \overline{\Delta W_1(t_1, T)\Delta W_2(t_2, T)}, \tag{20}$$

where

$$\Delta W_j(t_j, T) = W_j(t_j, T) - \overline{W_j(t_j, T)} \tag{21}$$

are the fluctuations of the integrated intensities of the light incident on the two detectors.

The formula (20) shows that *the correlation $\overline{\Delta n_1 \, \Delta n_2}$ between the fluctuations of the photocount at the two detectors is proportional to the correlation between the fluctuations in the integrated intensities of the classical field incident on the two detectors.*

If the integration time T of the detector is short relative to the coherence time $t_c = 2\pi/\Delta\omega$ of the incident light, we have from Eq. (3) that

$$W(t, T) \approx TI(t) \tag{22}$$

and Eq. (20) reduces to

$$\overline{\Delta n_1 \, \Delta n_2} = \alpha_1 \alpha_2 T^2 \overline{\Delta I_1(t_1)\Delta I_2(t_2)}, \tag{23}$$

where $\Delta I(t) = I(t) - \overline{I(t)}$ denotes the fluctuation of the intensity. Except for a slight change in notation, the right-hand side of Eq. (23) is identical with the left-hand side of Eq. (10) of Section 7.2 which, as we have seen, is the basic quantity from which angular diameters of stars are determined by intensity interferometry.

Equation (23) confirms that the correlation of intensity fluctuations of the classical field incident on the two detectors can indeed be determined from measurements of correlations of the fluctuations in the number of electrons emitted by the two illuminated photodetectors, in spite of the non-classical nature of the photoelectric effect.

Equation (20) contains the essence of the Hanbury Brown–Twiss effect. In practice it is, of course, also necessary to have an estimate of the signal-to-noise ratio. For discussion of that topic we refer the reader elsewhere.[1] It turns out that the signal-to-noise ratio depends

---

[1] See R. Hanbury Brown and R. Q. Twiss, *Proc. Roy. Soc.* (London), **A242** (1957), 300–324, Section 3(d). See also L. Mandel, in *Progress in Optics*, E. Wolf ed. (North-Holland, Amsterdam, 1963), Vol. II, Section 4.3, pp. 181–248.

rather significantly on the degeneracy parameter δ of the light [see Appendix I (c)] and that when δ is much smaller than unity it is essentially proportional to δ. We have shown in Appendix I that $\delta \ll 1$ for thermal sources at laboratory temperatures and for light from stellar sources. It is mainly for this reason that the effect had originally been so difficult to demonstrate. Since for laser light the degeneracy parameter is generally very large, of the order of 10^{17} or greater, one might expect that the effect would be very easy to demonstrate with laser light. However, as we have already noted, this is not so. A good-quality laser will generate a beam in which the intensity fluctuations ΔI are negligible. Consequently correlations in intensity fluctuations are essentially absent and the large value of the degeneracy parameter is then irrelevant. The effect can, however, easily be observed with so-called pseudo-thermal sources, which we mentioned in Section 7.3.

Finally we recall that in the preceding discussion we represented the incident field by a single scalar wave function. This is equivalent to assuming that the incident field is linearly polarized. For a discussion of intensity fluctuations and the photoelectric detection of light of some other states of polarization we refer the reader to a paper by Mandel.[1]

7.6 Determination of statistical properties of light from photocount measurements[2]

Let us now return to Mandel's formula (5) of Section 7.5.1 for the probability that n electrons are emitted in the time interval $(t, t + T)$ under the influence of incident light:

$$P(n, t, T) = \int_0^\infty \frac{[\alpha W(t, T)]^n}{n!} e^{-\alpha W(t,T)} p(W) dW. \tag{1}$$

Here

$$W(t, T) = \int_t^{t+T} I(t') dt' \tag{2}$$

represents the intensity of the incident light, integrated over the time interval $(t, t + T)$, and α is the photo-efficiency of the detector.

In principle, the photocount distribution $P(n, t, T)$ may be measured. We will now consider how one can derive the probability distribution $p(W)$ of the integrated intensity of the light incident on the detector from knowledge of that distribution. This amounts to finding the inverse of the Poisson transform which appears on the right-hand side of Eq. (1). For this purpose let us first introduce the function

$$F(x) = \int_0^\infty e^{ixW} p(W) e^{-\alpha W} dW. \tag{3}$$

[1] L. Mandel, *Proc. Phys. Soc.* (London), **81** (1963), 1104–1114.
[2] The analysis presented in this section is largely based on a paper by E. Wolf and C. L. Mehta, *Phys. Rev. Lett.* **13** (1964), 705–707.

Taking the Fourier inverse of $F(x)$ we have

$$p(W) = \frac{e^{\alpha W}}{2\pi} \int_{-\infty}^{\infty} F(x) e^{-ixW}\, dx. \tag{4}$$

Next let us expand the exponential on the right-hand side of Eq. (3) in a power series. This gives, formally at any rate,

$$F(x) = \int_0^{\infty} \sum_{n=0}^{\infty} \frac{(ixW)^n}{n!} p(W) e^{-\alpha W}\, dW$$

$$= \sum_{n=0}^{\infty} \frac{(ix)^n}{n!} \int_0^{\infty} W^n p(W) e^{-\alpha W}\, dW. \tag{5}$$

Now Mandel's formula (1) implies that

$$\int_0^{\infty} W^n p(W) e^{-\alpha W}\, dW = \frac{n!}{\alpha^n} P(n, t, T) \tag{6}$$

and hence Eq. (5) gives

$$F(x) = \sum_{n=0}^{\infty} \left(\frac{ix}{\alpha}\right)^n P(n, t, T). \tag{7}$$

It follows that the required probability distribution $p(W)$ may be obtained from knowledge of the photocount distribution $P(n, t, T)$ for $n = 0, 1, 2, \ldots$ by first evaluating the function $F(x)$, defined by Eq. (7), and then using Eq. (4).

Let us illustrate the inversion procedure by a simple example. Suppose that the photocount distribution $P(n, t, T)$ is Poissonian, i.e. that

$$P(n, t, T) = \frac{\bar{n}^n e^{-\bar{n}}}{n!}. \tag{8}$$

Equation (7) then gives

$$F(x) = \sum_{n=0}^{\infty} \left(\frac{ix}{\alpha}\right)^n \frac{\bar{n}^n e^{-\bar{n}}}{n!}$$

$$= e^{-\bar{n}} \sum_{n=0}^{\infty} \frac{1}{n!} \left(\frac{ix\bar{n}}{\alpha}\right)^n$$

$$= e^{\bar{n}[(ix/\alpha)-1]}. \tag{9}$$

On substituting from Eq. (9) into Eq. (4) we obtain for $p(W)$ the formula

$$p(W) = e^{\alpha W} e^{-\bar{n}} \frac{1}{2\pi} \int_{-\infty}^{\infty} e^{-ix(W-\bar{n}/\alpha)} \, dx. \tag{10}$$

The integral on the right of Eq. (10) is just $2\pi\delta(W - \bar{n}/\alpha)$, where δ is the Dirac delta function. Hence

$$p(W) = \delta(W - \bar{W}), \tag{11}$$

where

$$\bar{W} = \frac{\bar{n}}{\alpha}. \tag{12}$$

This result implies that when the observed photocount distribution is strictly Poissonian, i.e. given by Eq. (8), which represents pure shot noise, the intensity of the incident light is completely stabilized in the sense that it does not fluctuate at all and W has then the constant value $\overline{W} = \bar{n}/\alpha$.

PROBLEMS

7.1 $V(\mathbf{r}, t)$ is an analytic signal representation of a linearly polarized, stationary, thermal field and

$$\Delta I(\mathbf{r}, t) = I(\mathbf{r}, t) - \langle I(\mathbf{r}, t) \rangle$$

represents the fluctuation of the instantaneous intensity of the field at the point $\mathbf{r}$. Show that one may determine the absolute value of the degree of coherence $\gamma(\mathbf{r}_1, \mathbf{r}_2, \tau)$ of the light at two points in the field from knowledge of the averaged intensities and of the correlation between the intensity fluctuations at the two points.

7.2 $\langle E \rangle$ represents the average energy in the frequency range $(\nu, \nu + d\nu)$ of blackbody radiation, at temperature T, contained in a volume V. It is known that
(i) when $h\nu/(k_B T) \ll 1$, $\langle E \rangle$ is given by the Rayleigh–Jeans law

$$\langle E \rangle = Z k_B T;$$

(ii) when $h\nu/(k_B T) \gg 1$, $\langle E \rangle$ is given by Wien's law

$$\langle E \rangle = Z h\nu e^{-h\nu/(k_B T)},$$

where

$$Z = \frac{8\pi V \nu^2 \, d\nu}{c^3}.$$

(a) Show that the variance of the energy fluctuations is given by the expressions

$$\langle (\Delta E)^2 \rangle = \begin{cases} \frac{1}{2}\langle E \rangle^2 & \text{when } h\nu/(k_B T) \ll 1, \\ h\nu\langle E \rangle & \text{when } h\nu/(k_B T) \gg 1. \end{cases}$$

(b) Assuming that the fluctuations in these two limiting cases are due to two different causes and that, in general, both are effective and give rise to fluctuations whose variance is the sum of the two variances stated in (a), determine, with the help of the Einstein–Fowler formula, the general expression for $\langle E \rangle$.

7.3 (a) Determine the mean and the variance of the probability distribution

$$p_1(n) = \frac{\mu^n}{(1 + \mu)^{n+1}},$$

where μ is a positive constant and the random variable n takes on all non-negative integer values.

(b) The probability distribution for the number of photons of energy $h\nu$ in a mode of blackbody radiation at temperature T may be shown to be given by the expression

$$p_2(n) = (1 - e^{-h\nu/(k_B T)})\, e^{-nh\nu/(k_B T)},$$

where k_B is the Boltzmann constant. Show that $p_2(n)$ may be expressed in the form $p_1(n)$ with an appropriate choice of the parameter μ.
Determine also the variance of $p_2(n)$ in the two limiting cases when

$$h\nu/(k_B T) \ll 1 \quad \text{and} \quad h\nu/(k_B T) \gg 1.$$

7.4 The time-dependent intensity of a single-mode amplitude-stabilized laser beam, incident normally on a photodetector, has the form

$$I(t) = \frac{1}{2} I_0 [1 + \cos(\omega_0 t + \theta)],$$

where I_0 and ω_0 are constants and θ is a random variable which is uniformly distributed on the interval $(0, \pi)$. Determine the mean and the variance of the number of photoelectrons emitted by the photodetector in a time interval of duration T.

7.5 A laser beam of constant average intensity is incident normally on the surface of a photodetector. What is the probability $p(n)$ that n photoelectrons will be emitted from the detector in the time interval $(t, t + T)$? Comment on the physical significance of the result.

7.6 A stochastic electromagnetic beam is incident normally on a photoelectric detector. It
is found that the probability of n photoelectrons being emitted in a time interval $(t, t + T)$
is the Bose–Einstein distribution for a cell in phase-space, viz.,

$$P(n) = \frac{\bar{n}^n}{(\bar{n} + 1)^{n+1}} .$$

Determine the probability $p(W)$ of the integrated intensity

$$W = \int_t^{t+T} I(t')dt'$$

of the incident light.
What is the corresponding result when $P(n)$ is the Poisson distribution

$$P(n) = \frac{\bar{n}^n e^{-\bar{n}}}{n!} ?$$

8

Elementary theory of polarization of stochastic electromagnetic beams

Up to now we have simplified our discussion of optical fields by treating them as scalar fields, i.e. we have ignored their polarization properties which arise from the vector nature of the electromagnetic field. In this and the next chapter we will discuss the extension of the theory by taking the vector nature into account for a broad and a useful class of vectors fields, namely stochastic electromagnetic beams.

8.1 The 2 × 2 equal-time correlation matrix of a quasi-monochromatic electromagnetic beam

Let us consider an electromagnetic beam, i.e. an electromagnetic field which propagates close to a particular direction, which we will take to be the z direction (see Fig. 8.1). We will assume that the fluctuations of the electric and the magnetic field vectors are characterized by ensembles which are statistically stationary, at least in the wide sense. Let $E_x(t)$ and $E_y(t)$ be the components, represented by complex analytic signals (see Section 2.3), of the electric field at some point P in the beam, in two mutually orthogonal directions, perpendicular to the z direction.

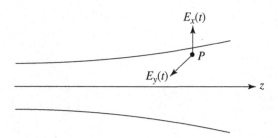

Fig. 8.1 Notation relating to propagation of an electromagnetic beam close to the z direction.

154

Assuming that the beam is quasi-monochromatic and of mean frequency $\bar{\omega}$, we may express E_x and E_y in the form

$$E_x(t) = a_1(t)e^{i[\phi_1(t)-\bar{\omega}t]}, \qquad E_y(t) = a_2(t)e^{i[\phi_2(t)-\bar{\omega}t]}. \tag{1}$$

As we learned in Section 2.3 in connection with the envelope representation of quasi-monochromatic signals, the amplitudes $a_1(t)$ and $a_2(t)$ and the phase functions $\phi_1(t)$ and $\phi_2(t)$ change slowly with time over intervals which are short relative to the coherence time $t_c \sim 2\pi/\Delta\omega$, $\Delta\omega$ being the effective bandwidth of the light.

The second-order correlation properties of the electric field at a point O may be characterized by a 2 × 2 correlation matrix

$$\mathbf{J} = \begin{bmatrix} \langle E_x^*(t)E_x(t)\rangle & \langle E_x^*(t)E_y(t)\rangle \\ \langle E_y^*(t)E_x(t)\rangle & \langle E_y^*(t)E_y(t)\rangle \end{bmatrix}, \tag{2}$$

the asterisk denoting the complex conjugate. The matrix $\mathbf{J}$ is called the *polarization matrix*. (A less appropriate name, "coherency matrix," is used in older literature.) This is an *equal-time* correlation matrix. Its diagonal elements are averages of the intensities associated with each of the two components and its off-diagonal elements are analogous to the mutual intensity $J(\mathbf{r}_1, \mathbf{r}_2)$ of a scalar field, which we encountered in Section 3.1. However, whilst the mutual intensity of the scalar field $V(\mathbf{r}, t)$ is a measure of correlations, at the same instant of time, at two points $\mathbf{r}_1$ and $\mathbf{r}_2$, the off-diagonal elements of this matrix are measures of the "equal-time" correlations between the mutually orthogonal components E_x and E_y of the electric field at a particular point.

We saw in Section 3.1 that the mutual intensity, and hence the diagonal elements of the matrix, may be determined by means of Young's interference experiment. We will now show that also the off-diagonal elements of the polarization matrix $\mathbf{J}$ may be determined by relatively simple interference experiments.

Suppose that the beam is passed through a compensator and then through a polarizer (Fig. 8.2).

Let ε_1 and ε_2 be the phase changes introduced in the components $E_x(t)$ and $E_y(t)$ by the compensator as the beam propagates from a plane $z = z_0$ to a plane $z = z_1$ and let θ denote

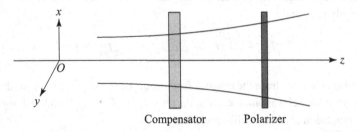

Compensator   Polarizer

Fig. 8.2 Transmission of a quasi-monochromatic beam through a system consisting of a compensator and a polarizer.

the angle which the directions of vibrations of the electric vector emerging from the polarizer makes with the x-axis. Apart from an unessential constant phase factor and ignoring reflection losses at the surfaces of the compensator and the polarizer, the linearly polarized electric field vector which emerges from the system is given by the expression

$$\mathbf{E}(t, \varepsilon_1, \varepsilon_2, \theta) = [E_x(t)e^{i\varepsilon_1} \cos\theta + E_y(t)e^{i\varepsilon_2} \sin\theta]\mathbf{i}_\theta, \tag{3}$$

where $\mathbf{i}_\theta$ is the unit vector in the direction of the linearly polarized electric vector, i.e. the two-dimensional unit vector with components $(\cos\theta, \sin\theta)$.

The averaged intensity, or, more precisely, the expectation value of the electric energy density of the light transmitted by the system, is given, in suitable units, by the formula

$$I(\varepsilon_1, \varepsilon_2, \theta) = \langle \mathbf{E}^*(t, \varepsilon_1, \varepsilon_2, \theta) \cdot \mathbf{E}(t, \varepsilon_1, \varepsilon_2, \theta) \rangle, \tag{4}$$

apart from a proportionality constant which depends on the choice of units. On substituting from Eq. (3) into Eq. (4) we find that

$$I(\varepsilon_1, \varepsilon_2, \theta) \equiv I(\delta, \theta)$$
$$= J_{xx} \cos^2\theta + J_{yy} \sin^2\theta + J_{xy}e^{i\delta} \sin\theta\cos\theta + J_{yx}e^{-i\delta} \cos\theta\sin\theta, \tag{5}$$

where

$$\delta = \varepsilon_2 - \varepsilon_1 \tag{6}$$

and

$$
\begin{aligned}
J_{xx} &= \langle E_x^*(t)E_x(t) \rangle, & J_{xy} &= \langle E_x^*(t)E_y(t) \rangle, \\
J_{yx} &= \langle E_y^*(t)E_x(t) \rangle, & J_{yy} &= \langle E_y^*(t)E_y(t) \rangle,
\end{aligned}
\tag{7}
$$

are the elements of the polarization matrix $\mathbf{J}$. We note that

$$J_{yx} = J_{xy}^*, \tag{8}$$

i.e. the matrix is Hermitian. Moreover, it is also *non-negative definite*, i.e., with any (real or complex) numbers c_1 and c_2,

$$c_1^* c_1 J_{xx} + c_2^* c_2 J_{yy} + c_1^* c_2 J_{xy} + c_1 c_2^* J_{yx} \geq 0. \tag{9}$$

This result follows at once from the obvious fact that $\langle |c_1 E_x + c_2 E_y|^2 \rangle \geq 0$. From this inequality or by applying the Schwarz inequality to the term $J_{xy} = \langle E_x^*(t)E_y(t) \rangle$ and making use of the Hermiticity condition (8) one readily finds that

$$|J_{xy}| \leq \sqrt{J_{xx}} \sqrt{J_{yy}}. \tag{10}$$

We may characterize the correlation between the x and y components of the electric field by the *correlation coefficient*

$$j_{xy} = \frac{J_{xy}}{\sqrt{J_{xx}} \sqrt{J_{yy}}}. \tag{11}$$

The inequality (10) then implies that

$$0 \le |j_{xy}| \le 1. \tag{12}$$

The extreme value $|j_{xy}| = 1$ represents complete correlation; the other extreme value, $|j_{xy}| = 0$, represents complete absence of correlation between the fluctuating x and y components of the electric field vector.

If one measures the average intensity with various phase delays δ and angles θ of orientation by the use of phase plates and polarizers one obtains, using Eq. (5), a set of linear equations from which the four elements of the 2 × 2 polarization matrix **J** can be determined.[1]

Instead of characterizing the properties of the fluctuating electric field by the 2 × 2 polarization matrix **J**, one frequently uses an older representation, in terms of *Stokes parameters*. They may be defined in terms of averages involving the amplitudes and the phases of the x and y components of the complex electric field by the formulas

$$\left.\begin{aligned}
s_0 &= \langle a_1^2(t) \rangle + \langle a_2^2(t) \rangle, \\
s_1 &= \langle a_1^2(t) \rangle - \langle a_2^2(t) \rangle, \\
s_2 &= 2\langle a_1(t) a_2(t) \cos[\phi_1(t) - \phi_2(t)] \rangle, \\
s_3 &= 2\langle a_1(t) a_2(t) \sin[\phi_1(t) - \phi_2(t)] \rangle.
\end{aligned}\right\} \tag{13}$$

It follows from Eqs. (13), (7) and (1) that the Stokes parameters and the elements of the polarization matrix **J** are related by the formulas

$$\left.\begin{aligned}
s_0 &= J_{xx} + J_{yy}, \\
s_1 &= J_{xx} - J_{yy}, \\
s_2 &= J_{xy} + J_{yx}, \\
s_3 &= i(J_{yx} - J_{xy})
\end{aligned}\right\} \tag{14a}$$

and

$$\left.\begin{aligned}
J_{xx} &= \tfrac{1}{2}(s_0 + s_1), \\
J_{yy} &= \tfrac{1}{2}(s_0 - s_1), \\
J_{xy} &= \tfrac{1}{2}(s_2 + is_3), \\
J_{yx} &= \tfrac{1}{2}(s_2 - is_3).
\end{aligned}\right\} \tag{14b}$$

---

[1] Expressions for the four elements in terms of a particularly convenient choice of the parameters δ and θ are given in B&W, p. 621.

One can readily derive formulas that describe the changes which the polarization matrix $\mathbf{J}$ or the so-called *Stokes vector* $\mathbf{s} \equiv (s_0, s_1, s_2, s_3)$ undergoes as the beam passes through various linear non-imaging devices such as polarizers, compensators and rotators. In the correlation-matrix formalism they are represented by 2×2 "transmission" matrices. In the formalism which uses the Stokes parameters they are represented by 4×4 matrices known as Mueller matrices. Since this topic is treated in many publications,[1] we will not discuss it here.

When the electromagnetic beam is monochromatic, the Stokes parameters are particularly useful for representing certain geometrical properties of the oscillating field vector, as we will see in Section 8.2.4.

8.2 Polarized, unpolarized and partially polarized light. The degree of polarization

Of special interest are fields for which the modulus of the degree of correlation j_{xy}, defined by Eq. (11) of Section 8.1, takes on one of the extreme values, either $|j_{xy}| = 1$ or $j_{xy} = 0$. We will now consider these two cases.

8.2.1 Completely polarized light

Suppose that

$$|j_{xy}| = 1. \tag{1}$$

In this case the x and the y components of the electric field are completely correlated. It follows at once from Eqs. (11) and (8) of Section 8.1 that then

$$\text{Det}\,\mathbf{J} \equiv J_{xx}J_{yy} - J_{xy}J_{yx}$$

$$= 0. \tag{2}$$

Conversely, the vanishing of the determinant of the polarization matrix may readily be seen to imply Eq. (1), i.e. it implies complete correlation between the x and the y components of the electric vector.

It is known from elementary theory of matrices that the determinant of the polarization matrix $\mathbf{J}$ is invariant with respect to rotation of the x- and the y-axes about the direction of propagation of the beam. Hence if the x and the y components of the electric field are completely correlated for a particular set of coordinate axes they will also be completely correlated for any such pair of axes.

---

[1] See, for example, G. E. Parrent and P. Roman, *Nuovo Cimento* **15** (1960), 370–388; E. L. O'Neill, *Introduction to Statistical Optics* (Addison-Wesley, Reading, MA (1963); reprinted by Dover, New York, 2004); and E. Collett, *Polarized Light* (Marcel Dekker, New York, 1993), Chapter 5.

From Eq. (1) and from Eq. (11) of Section 8.1, making use of the Hermiticity of the polarization matrix expressed by Eq. (8) of Section 8.1, it follows that

$$
\mathbf{J} = \begin{bmatrix} J_{xx} & \sqrt{J_{xx}}\sqrt{J_{yy}}\,e^{i\alpha} \\ \sqrt{J_{xx}}\sqrt{J_{yy}}\,e^{-i\alpha} & J_{yy} \end{bmatrix},
\tag{3}
$$

where α is real.

Equation (1), or, equivalently, the statement that E_x and E_y are completely correlated, characterizes light which is said to be *completely polarized* at the point $P(\mathbf{r})$. This terminology is used because of a formal analogy between this situation and the behavior of a (necessarily deterministic) monochromatic field. Such fields will be discussed in Section 8.2.4. Here we only explain the reason for the analogy.

A monochromatic electromagnetic beam is, at every point, completely polarized in the sense that with increasing time the end point of its electric field vector moves on an ellipse (see B&W, Section 1.4.3). One says that the field is *elliptically polarized*. In some cases the ellipse may, of course, reduce to a circle or a straight line, in which cases we speak of *circular* and *linear polarization*, respectively.

Consider a monochromatic beam propagating close to the positive z direction. Let

$$
E_x(z, t) = \mathbf{e}_1 e^{i(kz-\omega t)}, \qquad E_y(z, t) = \mathbf{e}_2 e^{i(kz-\omega t)}
\tag{4}
$$

($k = \omega/c$, c being the speed of light in vacuum) be the components of the (complex) electric vector at an arbitrary point, along two mutually orthogonal directions, specified by unit vectors $\mathbf{e}_1$ and $\mathbf{e}_2$, perpendicular to the direction of propagation. One can associate with this monochromatic wave a 2×2 matrix which is somewhat analogous to the polarization matrix [Eq. (2) of Section 8.1] of a random and, therefore, necessarily not monochromatic field, namely the matrix

$$
\mathbf{J} = \begin{bmatrix} E_x^* E_x & E_x^* E_y \\ E_y^* E_x & E_y^* E_y \end{bmatrix}
\tag{5a}
$$

$$
= \begin{bmatrix} e_1^* e_1 & e_1^* e_2 \\ e_2^* e_1 & e_2^* e_2 \end{bmatrix}.
\tag{5b}
$$

It is to be noted that the elements of this matrix do not involve any averaging.

It is evident that the determinant of this matrix has the value zero, just as is the case when the polarization matrix represents a completely polarized beam [see Eq. (2)]. This result implies a certain equivalence theorem, namely that the "canonical" experiment which employs only a compensator and a polarizer for determining the elements of the polarization matrix $\mathbf{J}$ cannot distinguish between a random quasi-monochromatic beam whose polarization matrix has the form given by Eq. (3) and a strictly monochromatic beam whose polarization matrix has the form (5b), with

$$
e_1 = \sqrt{J_{xx}}\,e^{i\alpha_1}, \qquad e_2 = \sqrt{J_{yy}}\,e^{i\alpha_2},
\tag{6}
$$

where α_1 and α_2 are real arbitrary constants.

8.2.2 Natural (unpolarized) light

We now consider the other extreme case, namely when the degree of correlation takes on the value zero, i.e. when

$$j_{xy} = 0, \tag{7}$$

irrespective of the particular choice of the x- and y-axes. Light with these properties is said to be natural light, because it very frequently occurs in nature, for example light reaching the Earth from most astronomical sources. For reasons that will be explained shortly, such light may also be said to be *unpolarized*.

Let us consider some implications of Eq. (7). For this purpose let us first examine how the matrix **J** changes when the axes are rotated by an angle, say Θ, in the anticlockwise sense around the direction of propagation of the beam. If $E_{x'}$, $E_{y'}$ are the components of **E** referred to the rotated coordinate system $O_{x'}$, $O_{y'}$ (see Fig. 8.3) one has

$$\left. \begin{aligned} E_{x'} &= E_x \cos \Theta + E_y \sin \Theta, \\ E_{y'} &= -E_x \sin \Theta + E_y \cos \Theta. \end{aligned} \right\} \tag{8}$$

The elements of the polarization matrix, referred to the rotated coordinate system, are $J_{k'\varrho'} = \langle E_{k'}^* E_{\varrho'} \rangle$ and, using Eq. (8), we see that the polarization matrix in the new coordinate system is

$$\mathbf{J}' \equiv \begin{bmatrix} J_{xx}c^2 + J_{yy}s^2 + (J_{xy} + J_{yx})cs & (J_{yy} - J_{xx})cs + J_{xy}c^2 - J_{yx}s^2 \\ (J_{yy} - J_{xx})cs + J_{yx}c^2 - J_{xy}s^2 & J_{xx}s^2 + J_{yy}c^2 - (J_{xy} + J_{yx})cs \end{bmatrix}, \tag{9}$$

where $c = \cos \theta$ and $s = \sin \theta$.

According to the definition of the correlation coefficient [Eq. (11) of Section 8.1] and using the Hermiticity condition [Eq. (8) of that section], the requirement (7) implies that for natural light

$$J_{xy} = J_{yx} = 0, \tag{10}$$

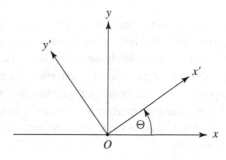

Fig. 8.3 Notation relating to the change in the correlation matrix **J** when the axes are rotated about the direction of propagation of the beam.

irrespective of the particular choice of the x'- and y'-axes. In view of the transformation law given by Eq. (9), Eq. (10) implies that with any choice of θ the term $(J_{yy} - J_{xx})\cos\theta\sin\theta = 0$, i.e. that with any choice of the x- and y-axes

$$J_{yy} = J_{xx}. \tag{11}$$

Using Eqs. (10) and (11) it follows that the *polarization matrix of natural light* has the form

$$\mathbf{J} = J_{xx}\begin{bmatrix} 1 & 0 \\ 0 & 1 \end{bmatrix}, \tag{12}$$

i.e. $\mathbf{J}$ *is proportional to the unit matrix*. The proportionality factor J_{xx} is also independent of the choice of the axes, because, when Eqs. (10) and (11) hold, the transformed polarization matrix is independent of θ. For reasons that will become apparent shortly such a beam is also said to be *unpolarized*.

Let us consider the transmission of natural light through the system shown in Fig. 8.2, consisting of a compensator, which introduces a phase delay δ, and a polarizer, which transmits the component of the electric field vector which makes the angle θ with the x-axis. The intensity $I(\delta, \theta) \equiv I(\varepsilon_1, \varepsilon_2, \theta)$ of the light emerging from such a system is given by Eq. (5) of Section 8.1, viz.,

$$I(\delta, \theta) = J_{xx}\cos^2\theta + J_{yy}\sin^2\theta + J_{xy}e^{i\delta}\sin\theta\cos\theta + J_{yx}e^{-i\delta}\cos\theta\sin\theta. \tag{13}$$

On substituting from Eqs. (10) and (11) into this formula we obtain for the intensity of the transmitted beam the expression

$$I(\delta, \theta) = J_{xx}, \tag{14}$$

i.e. only half of the intensity of the incident wave is transmitted. The formula shows that, when a beam of natural light is transmitted by a compensator and a polarizer, the intensity of the light emerging from this system is unaffected both by the retardation introduced by the compensator and by the orientation of the polarizer.

8.2.3 Partially polarized light and the degree of polarization

We considered two extreme situations: a light beam for which the absolute value of the correlation coefficient j_{xy} has the extreme value unity and a light beam for which it has the other extreme value, zero. We will now show that the polarization matrix of any light beam may be, at each point, expressed *uniquely* as the sum of these two kinds of beams.

The ratio of the intensity of the polarized part to the total intensity is called the *degree of polarization*. The nature of the polarized portion (i.e. the length of the principal axes of its polarization ellipse and the orientation of the ellipse) together with the degree of polarization are said to represent the *state of polarization* of the beam. It generally changes as the beam propagates[1] or is scattered, for example, by a solid body or by a system of particles.

[1] That the degree of polarization may change on propagation, even in free space, appears to have been first demonstrated by D. F. V. James, *J. Opt. Soc. Amer* **A11** (1994), 1641–1643.

Such changes are often of considerable practical interest because they provide information about the physical system that interacts with the beam.

Let us then consider the possibility of expressing the polarization matrix of a beam propagating close to the z direction in the form

$$\mathbf{J} = \mathbf{J}^{(\mathrm{p})} + \mathbf{J}^{(\mathrm{u})}, \tag{15}$$

where the two matrices on the right represent completely polarized light (superscript p) and completely unpolarized light (superscript u). According to Eqs. (13) and (12) the matrices are of the form

$$\mathbf{J}^{(\mathrm{p})} = \begin{bmatrix} B & D \\ D^* & C \end{bmatrix}, \qquad \mathbf{J}^{(\mathrm{u})} = A \begin{bmatrix} 1 & 0 \\ 0 & 1 \end{bmatrix}, \tag{16}$$

where

$$A \geq 0, \qquad B \geq 0, \qquad C \geq 0, \tag{17}$$

and

$$BC - DD^* = 0. \tag{18}$$

With $J_{k\ell}$, $(k = x, y, \ell = x, y)$, representing the elements of the polarization matrix $\mathbf{J}$, Eqs. (15) and (16) imply that

$$\left. \begin{aligned} A + B &= J_{xx}, & D &= J_{xy}, \\ D^* &= J_{yx}, & A + C &= J_{yy}. \end{aligned} \right\} \tag{19}$$

On substituting for B, C and D from Eq. (19) into Eq. (18) we obtain the following equation for the matrix element A:

$$(J_{xx} - A)(J_{yy} - A) - J_{xy}J_{yx} = 0. \tag{20}$$

This equation shows that A is an eigenvalue of the polarization matrix $\mathbf{J}$. A simple calculation shows that the solution to this equation may be expressed in the form

$$A = \frac{1}{2} \left\{ \mathrm{Tr}\, \mathbf{J} \pm [(\mathrm{Tr}\, \mathbf{J})^2 - 4\, \mathrm{Det}\, \mathbf{J}]^{\frac{1}{2}} \right\}, \tag{21}$$

where Tr denotes the trace and Det the determinant. Since according to Eqs. (8) and (9) of Section 8.1 the polarization matrix is Hermitian and non-negative definite, both eigenvalues A are necessarily non-negative, as may also be verified by direct calculation.

Let us consider first the root A given by Eq. (21), with the negative sign in front of the square root:

$$A = \frac{1}{2}(J_{xx} + J_{yy}) - \frac{1}{2}\sqrt{(J_{xx} + J_{yy})^2 - 4\, \mathrm{Det}\, \mathbf{J}}. \tag{22a}$$

On substituting this root into the diagonal terms in Eqs. (19) we obtain the following expressions for the matrix elements B and C:

$$B = \frac{1}{2}(J_{xx} - J_{yy}) + \frac{1}{2}\sqrt{(\text{Tr }\mathbf{J})^2 - 4\,\text{Det }\mathbf{J}}, \tag{22b}$$

$$C = \frac{1}{2}(J_{yy} - J_{xx}) + \frac{1}{2}\sqrt{(\text{Tr }\mathbf{J})^2 - 4\,\text{Det }\mathbf{J}}. \tag{22c}$$

If we make use of the Hermiticity relation $J_{yx} = J_{xy}^*$ [Eq. (8) of Section 8.1] we find at once that

$$\left[(\text{Tr }\mathbf{J})^2 - 4\,\text{Det }\mathbf{J}\right]^{\frac{1}{2}} = \sqrt{(J_{xx} - J_{yy})^2 + 4\,|J_{xy}|^2} \geq |J_{xx} - J_{yy}|. \tag{23}$$

Consequently the matrix elements B and C given by Eqs. (22) are necessarily non-negative as required by two of the inequalities in (17). On the other hand, the other expression for A given by Eq. (21), with the positive sign in front of the square root, may readily be shown to yield negative values for B and C and, therefore, does not satisfy the inequalities (17). Thus we have shown that there is a unique decomposition of the polarization matrix $\mathbf{J}$ in the form given by Eq. (15), subject to the constraints given by Eqs. (17) and (18). This result implies that *any statistically stationary light beam may, at each point, be regarded as being the sum of two beams, one of which is completely polarized and the other completely unpolarized.*

It follows from the expression for the polarization matrix $\mathbf{J}^{(p)}$ of the polarized part [the first matrix in Eq. (16)] and from Eqs. (22b) and (22c) that its trace $(B + C)$ is given by the expression

$$\text{Tr }\mathbf{J}^{(p)} = \sqrt{(\text{Tr }\mathbf{J})^2 - 4\,\text{Det }\mathbf{J}}. \tag{24}$$

The trace of the polarization matrix $\mathbf{J}^{(u)}$ of the unpolarized part is, according to Eqs. (16) and (21), given by the formula

$$\text{Tr }\mathbf{J}^{(u)} = \text{Tr }\mathbf{J} - \sqrt{(\text{Tr }\mathbf{J})^2 - 4\,\text{Det }\mathbf{J}}, \tag{25}$$

where we used the fact demonstrated earlier that only the expression for A with the negative sign in front of the square-root sign is admissible. Now, according to the definition of the polarization matrix [Eq. (2) of Section 8.1], the trace of that matrix is proportional to the average electric energy density and may, therefore, be regarded as a measure of the intensity, say I, of the light. Using Eqs. (24) and (25) it follows that the degree of polarization of light, defined as the ratio of the intensity of the polarized portion $I^{(p)}$ to the total intensity I at the point under consideration, is given by the expression

$$\mathcal{P} \equiv \frac{I^{(p)}}{I} = \sqrt{1 - \frac{4\,\text{Det }\mathbf{J}}{(\text{Tr }\mathbf{J})^2}}. \tag{26}$$

Since both the determinant and the trace of the correlation matrix are invariants under the rotation of the x- and y-axes about the z direction, the degree of polarization $\mathcal{P}$ is independent of any particular choice of the x- and y-axes. Also, in view of the inequality (23) one can readily deduce from Eq. (26) that

$$0 \le \mathcal{P} \le 1. \tag{27}$$

We see at once, on using Eq. (2) in Eq. (26), that, for light that we previously called completely polarized light, the degree of polarization $\mathcal{P}$, given by Eq. (26), does indeed have the extreme value $\mathcal{P} = 1$.

For natural light we have, from Eq. (12), $\mathrm{Tr}\,\mathbf{J} = 2J_{xx}$ and $\mathrm{Det}\,\mathbf{J} = J_{xx}^2$, so the degree of polarization given by Eq. (26) has, in this case, the other extreme value $\mathcal{P} = 0$, i.e. it may be said to be *unpolarized*.

When the degree of polarization has a value between these two extremes, i.e. when $0 < \mathcal{P} < 1$, we say that the light is *partially polarized*.

As we saw earlier the polarization matrix is Hermitian [Eq. (8) of Section 8.1]. Hence by application of a well-known theorem about such matrices[1] it can be diagonalized by a unitary transformation (which, however, need not be a rotation). It will then have the form

$$\mathbf{J}' = \begin{bmatrix} \lambda_1 & 0 \\ 0 & \lambda_2 \end{bmatrix}, \tag{28}$$

λ_1 and λ_2 being the eigenvalues of the polarization matrix. Moreover, because the polarization matrix is also non-negative definite [Eq. (9) of Section 8.1] the eigenvalues are necessarily non-negative. Since the trace and the determinant of a matrix are invariant under rotation of the x- and y-axes about the z direction,[2] the determinant of the original polarization matrix $\mathbf{J}$ and the determinant of its diagonalized matrix [Eq. (28)] must be equal to each other and the same is true of the traces. Hence

$$\mathrm{Det}\,\mathbf{J} = \lambda_1 \lambda_2, \tag{29a}$$

$$\mathrm{Tr}\,\mathbf{J} = \lambda_1 + \lambda_2, \tag{29b}$$

and the degree of polarization, given by Eq. (26), may, therefore, be expressed in the simple form

$$\mathcal{P} = \sqrt{1 - \frac{4\lambda_1 \lambda_2}{(\lambda_1 + \lambda_2)^2}}, \tag{30a}$$

i.e.

$$\mathcal{P} = \frac{|\lambda_1 - \lambda_2|}{\lambda_1 + \lambda_2}. \tag{30b}$$

---

[1] F. W. Byron and R. W. Fuller, *Mathematics of Classical and Quantum Physics* (Addison-Wesley, Reading, MA, 1969); reprinted by Dover, New York, 1992, Vol. 1, p. 165, Theorem 4.20.
[2] See, for example, F. W. Byron and R. W. Fuller, *loc. cit.*, p. 119, Theorem 3.13.

Since the eigenvalues are independent of the choice of the x- and y-axes it is clear that the degree of polarization $\mathcal{P}$ is also independent of that choice, as, indeed, must be the case if $\mathcal{P}$ is to have an unambiguous physical meaning.

8.2.4 The geometrical significance of complete polarization. The Stokes parameters of completely polarized light. The Poincaré sphere

We saw in Section 8.2.1 that completely polarized light is characterized by the property that the x and y components of the electric field vector are completely correlated and that, consequently, the polarization matrix is expressible in the form [Eq. (5b) of Section 8.2.1]

$$\mathbf{J} = \begin{bmatrix} e_1^* e_1 & e_1^* e_2 \\ e_2^* e_1 & e_2^* e_2 \end{bmatrix}, \tag{31}$$

where e_1 and e_2 are independent of time. We have also seen that such a matrix is indistinguishable from the polarization matrix of a monochromatic plane wave with $\omega = \bar{\omega}$ and $k = \bar{k}$,

$$E_x(z, t) = e_1 e^{i(kz - \omega t)}, \qquad E_y(z, t) = e_2 e^{i(kz - \omega t)}. \tag{32}$$

We will now consider some geometrical implications of Eqs. (32).

Equations (32) represent the Cartesian components of the electric field vector in complex form. The physically meaningful quantities are their real parts, i.e.

$$E_x^{(r)}(z, t) = |e_1| \cos(\alpha_1 + kz - \omega t), \tag{33a}$$

$$E_y^{(r)}(z, t) = |e_2| \cos(\alpha_2 + kz - \omega t), \tag{33b}$$

where α_1 and α_2 are the phases of e_1 and e_2, respectively.

It will be convenient to set

$$p_x = |e_1| \cos(\alpha_1 + kz), \qquad q_x = |e_1| \sin(\alpha_1 + kz), \tag{34a}$$

$$p_y = |e_2| \cos(\alpha_2 + kz), \qquad q_y = |e_2| \sin(\alpha_2 + kz). \tag{34b}$$

By expanding the cosine terms on the right-hand sides of Eqs. (33) and (34) by the use of elementary trigonometric identities, the components of the electric field may be expressed in the form

$$E_x^{(r)}(t) = p_x \cos(\omega t) + q_x \sin(\omega t), \tag{35a}$$

$$E_y^{(r)}(t) = p_y \cos(\omega t) + q_y \sin(\omega t), \tag{35b}$$

where we suppressed the explicit dependences of the various quantities on z. On combining the two scalar formulas (35) into a simple vector formula we have

$$\mathbf{E}^{(r)}(t) = \mathbf{p} \cos(\omega t) + \mathbf{q} \sin(\omega t), \tag{36}$$

where $\mathbf{p} \equiv (p_x, p_y)$ and $\mathbf{q} \equiv (q_x, q_y)$.

Equation (36) shows that, with increasing time, the end point of the electric field vector $\mathbf{E}^{(r)}(t)$ at a fixed point in space describes a curve in the plane containing the two real vectors $\mathbf{p}$ and $\mathbf{q}$ and passes through their end points. Because $\cos(\omega t)$ and $\sin(\bar{\omega}t)$ are periodic functions of time t, the curve is evidently closed. We will now show that, in general, the curve is an ellipse. To see this let us rewrite Eq. (36) as

$$\mathbf{E}^{(r)}(t) = \mathcal{R}e\{(\mathbf{p} + i\mathbf{q})e^{-i\omega t}\}, \tag{37}$$

where $\mathcal{R}e$ again denotes the real part. Let us set

$$(\mathbf{p} + i\mathbf{q}) = (\mathbf{a} + i\mathbf{b})e^{i\varepsilon}, \tag{38}$$

where ε is a scalar parameter. In terms of $\mathbf{p}$, $\mathbf{q}$ and ε one evidently has

$$\mathbf{a} = \mathbf{p} \cos \varepsilon + \mathbf{q} \sin \varepsilon, \tag{39a}$$

$$\mathbf{b} = -\mathbf{p} \sin \varepsilon + \mathbf{q} \cos \varepsilon. \tag{39b}$$

Let us now choose the parameter ε so that the vectors $\mathbf{a}$ and $\mathbf{b}$ are mutually orthogonal with $|\mathbf{a}| \geq |\mathbf{b}|$. In that case ε must evidently satisfy the equation

$$(\mathbf{p} \cos \varepsilon + \mathbf{q} \sin \varepsilon) \cdot (-\mathbf{p} \sin \varepsilon + \mathbf{q} \cos \varepsilon) = 0. \tag{40}$$

From this equation it readily follows that

$$\tan(2\varepsilon) = \frac{2\mathbf{p} \cdot \mathbf{q}}{\mathbf{p}^2 - \mathbf{q}^2}. \tag{41}$$

We may take as the parameters which specify the behavior of the electric field at any particular point the time-independent components of the mutually orthogonal vectors $\mathbf{a}$ and $\mathbf{b}$ and the associated parameter ε, rather than the six Cartesian components of the vectors $\mathbf{p}$ and $\mathbf{q}$. From Eqs. (37) and (38) it then follows that

$$\mathbf{E}^{(r)}(t) = \mathbf{a} \cos(\omega t - \varepsilon) + \mathbf{b} \sin(\omega t - \varepsilon). \tag{42}$$

If we take Cartesian axes with the origin at the point where the electric field is being considered and with the x and y directions along the vectors $\mathbf{a}$ and $\mathbf{b}$, respectively, we evidently have

$$E_x^{(r)} = a \cos(\omega t - \varepsilon), \qquad E_y^{(r)} = b \sin(\omega t - \varepsilon), \qquad E_z = 0 \tag{43}$$

($a = |\mathbf{a}|$, $b = |\mathbf{b}|$). These equations represent an ellipse in the x, y-plane, known as the *polarization ellipse*,

$$\frac{\left[E_x^{(r)}\right]^2}{a^2} + \frac{\left[E_y^{(r)}\right]^2}{b^2} = 1, \tag{44}$$

with the semi-axes a and b along the x and y coordinate axes. By elementary geometry it may be shown that $\mathbf{p}$ and $\mathbf{q}$ are a pair of conjugate semi-diameters of the ellipse.[1]

The ellipse may be traversed in one of two possible senses. In conformity with frequently used terminology we say that the polarization is *right-handed* when, to an observer looking in the direction from which the light is arriving, the end point of the electric field vector would appear to traverse the ellipse in the clockwise sense. In the opposite case we say that the polarization is *left-handed*. These two situations are distinguished by the sign of the scalar triple product $[\mathbf{a}, \mathbf{b}, \nabla\varepsilon] = [\mathbf{p}, \mathbf{q}, \nabla\varepsilon]$.

We may readily determine the semi-axes a and b of the polarization ellipse in terms of the original quantities $\mathbf{p}$ and $\mathbf{q}$ and the parameter ε, given by Eq. (41). We have from (39a)

$$
\begin{aligned}
a^2 &= p^2 \cos^2\varepsilon + q^2 \sin^2\varepsilon + 2\mathbf{p}\cdot\mathbf{q}\cos\varepsilon\sin\varepsilon \\
&= \frac{1}{2}(p^2 + q^2) + \frac{1}{2}(p^2 - q^2)\cos(2\varepsilon) + \mathbf{p}\cdot\mathbf{q}\sin(2\varepsilon).
\end{aligned}
\tag{45}
$$

From Eq. (41) it follows that

$$
\sin(2\varepsilon) = \frac{2\mathbf{p}\cdot\mathbf{q}}{\sqrt{(p^2 - q^2)^2 + 4(\mathbf{p}\cdot\mathbf{q})^2}}, \qquad
\cos(2\varepsilon) = \frac{p^2 - q^2}{\sqrt{(p^2 - q^2)^2 + 4(\mathbf{p}\cdot\mathbf{q})^2}}.
\tag{46}
$$

Hence

$$
a^2 = \frac{1}{2}\left[p^2 + q^2 + \sqrt{(p^2 - q^2)^2 + 4(\mathbf{p}\cdot\mathbf{q})^2} \right].
\tag{47a}
$$

Similarly, one finds that

$$
b^2 = \frac{1}{2}\left[p^2 + q^2 - \sqrt{(p^2 - q^2)^2 + 4(\mathbf{p}\cdot\mathbf{q})^2} \right].
\tag{47b}
$$

To determine the angle between $\mathbf{a}$ and $\mathbf{p}$ we express the equation of the polarization ellipse in a parametric form,

$$
E_x^{(\mathrm{r})} = a\cos\phi, \qquad E_y^{(\mathrm{r})} = b\sin\phi,
\tag{48}
$$

where ϕ is the so-called eccentricity angle (see Fig. 8.4). According to elementary geometry the angle ϕ is related to the polar angle θ of the point $(E_x^{(\mathrm{r})}, E_y^{(\mathrm{r})})$ by the formula

$$
\tan\theta = \frac{b}{a}\tan\phi.
\tag{49}
$$

---

[1] For discussion of conjugate diameters see, for example, A. Robinson, *An Introduction to Analytical Geometry* (Cambridge University Press, Cambridge, 1940), Vol. I, Section 14.7.

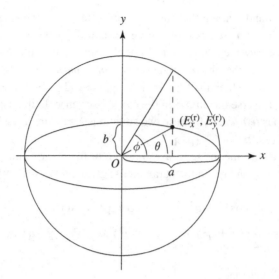

Fig. 8.4 Illustrating the meaning of the eccentricity angle ϕ of a point $(E_x^{(r)}, E_y^{(r)})$ on the polarization ellipse.

Comparison of Eqs. (43) and (48) shows that in the present case $\phi = \omega t - \varepsilon$. Now, according to Eq. (36), $\mathbf{E}^{(r)}(t) = \mathbf{p}$ when $t = 0$, so that the eccentricity angle of $\mathbf{p}$ is $-\varepsilon$. Equation (49) then implies that the angle ψ between $\mathbf{a}$ and $\mathbf{p}$ is given by the formula

$$\tan \psi = \frac{b}{a} \tan \varepsilon. \qquad (50)$$

Let us introduce an auxiliary angle β defined by the equation

$$\frac{q}{p} \equiv \tan \beta. \qquad (51)$$

It then follows from the formula (41) and from elementary trigonometric identities that

$$\tan(2\varepsilon) = \tan(2\beta)\cos \gamma, \qquad (52)$$

where γ is the angle between the vectors $\mathbf{p}$ and $\mathbf{q}$.

Let us summarize the results which we have just derived. If the real vectors $\mathbf{p}$ and $\mathbf{q}$ are given [see Eq. (36)], and if γ denotes the angle between them and β denotes the auxiliary angle defined by Eq. (51), then the semi-axes of the ellipse and the angle ψ which its major axis makes with the vector $\mathbf{p}$ are given by the formulas (47) and (50) (see also Fig. 8.5), where ε is given by Eqs. (52) and (41).

Two special cases are of interest, namely when the ellipse degenerates into a circle and when it degenerates into a straight line. In the first case one says that the electric field at the

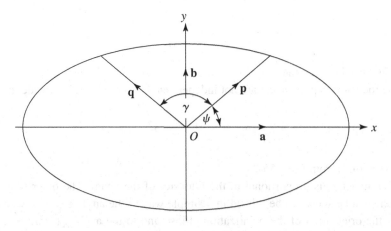

Fig. 8.5 Illustrating the significance of various parameters used for specifying the polarization ellipse and some useful relations between them:

$$\mathbf{E}^{(r)}(t) = \mathbf{p}\cos(\omega t) + \mathbf{q}\sin(\omega t)$$
$$= \mathbf{a}\cos(\omega t - \varepsilon) + \mathbf{b}\sin(\omega t - \varepsilon),$$
$$\tan\psi = \frac{b}{a}\tan\varepsilon,$$
$$\tan\beta = \frac{q}{p},$$
$$\tan(2\varepsilon) = \frac{2\mathbf{p}\cdot\mathbf{q}}{p^2 - q^2} = \tan(2\beta)\cos\gamma.$$

point under consideration is *circularly polarized*. Then $\mathbf{a}$ and $\mathbf{b}$ and, consequently, also ε are undetermined. According to Eq. (41), one then has

$$\mathbf{p}\cdot\mathbf{q} = p^2 - q^2 = 0. \tag{53}$$

When the ellipse degenerates into a straight line, i.e. when the wave is *linearly polarized*, there is no minor axis ($b^2 = 0$) and the formula (47b) then gives

$$(\mathbf{p}\cdot\mathbf{q})^2 = p^2 q^2. \tag{54}$$

We briefly mentioned towards the end of Section 8.1 an older representation of electromagnetic beams in terms of the so-called Stokes parameters [Eqs. (13) of that section]. In the special case when the beam is completely polarized the Stokes parameters are given by the simple expressions

$$\left.\begin{aligned}
s_0 &= |e_1|^2 + |e_2|^2, \\
s_1 &= |e_1|^2 - |e_2|^2, \\
s_2 &= 2|e_1||e_2|\cos\delta, \\
s_3 &= 2|e_1||e_2|\sin\delta,
\end{aligned}\right\} \tag{55}$$

where

$$\delta = \alpha_2 - \alpha_1 \tag{56}$$

is the difference between the phases α_1 and α_2 of the (generally complex) quantities e_1 and e_2. These four Stokes parameters are not independent but are related by the identity

$$s_0^2 = s_1^2 + s_2^2 + s_3^2, \tag{57}$$

as follows at once from Eqs. (55).

The parameter s_0 is proportional to the intensity of the beam. The other three Stokes parameters can be shown to be related in a simple way to the angle $\psi (0 \leq \psi \leq \pi)$ which specifies the orientation of the polarization ellipse and to the angle $\chi (-\pi/2 \leq \chi \leq \pi/2)$ which characterizes the ellipticity and the sense in which the ellipse is traversed. The following expressions for the Stokes parameters s_1, s_2, s_3 hold:

$$s_1 = s_0 \cos(2\chi)\cos(2\psi), \tag{58a}$$

$$s_2 = s_0 \cos(2\chi)\sin(2\psi), \tag{58b}$$

$$s_3 = s_0 \sin(2\chi). \tag{58c}$$

The derivation of these relations is somewhat lengthy. It is given elsewhere (see, for example, B&W, Section 1.4.2).

Equations (58) indicate a simple geometrical representation of all possible states of a polarized field. Evidently the three Stokes parameters [Eq. (58)] may be regarded as Cartesian coordinates of a point P on a sphere Σ of radius s_0, with 2χ and 2ψ being the spherical angular coordinates of that point (see Fig. 8.6). *Hence to every possible state of a fully polarized beam of intensity s_0 at an arbitrary point there corresponds a point P on the sphere Σ (known as the Poincaré sphere) and vice versa.* Since χ is positive or negative, respectively, according to whether the polarization is right-handed or left-handed, it follows from Eq. (58c) that a right-handed polarized state is represented by points on Σ which lie above the equatorial plane (the x, y-plane) and left-handed polarized states are represented by points which lie below that plane. Further, for linearly polarized light the phase difference δ defined by Eq. (56) is zero or is an integral multiple of π. According to the fourth equation of the expressions in Eq. (55), the Stokes parameter s_3 is then zero. Hence, linear polarization is represented by points in the equatorial plane. For circular polarization $|e_1| = |e_2|$ and $\delta = \pi/2$ or $-\pi/2$, according to whether the polarization is right- or left-handed; hence right-handed polarization is represented by the north pole ($s_1 = s_2 = 0$, $s_3 = s_0$) and left-handed circular polarization by the south pole ($s_1 = s_2 = 0$, $s_3 = -s_0$). This geometrical representation of different states of completely polarized light by points on a sphere is due to Poincaré. It is particularly useful in crystal optics for determining the changes in the state of polarization of light which traverses a crystalline medium.

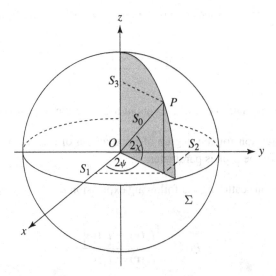

Fig. 8.6 The Poincaré sphere. Every point P on this sphere, of radius s_0, with Cartesian coordinates given by the Stokes parameters s_1, s_2, s_3, represents the state of polarization at a point in space. The angles 2χ and 2ψ are the polar angles specifying the direction of the vector $\overrightarrow{OP}$ [see Eqs. (58)]. Points above the equatorial plane represent right-handed polarization; points below it represent left-handed polarization. Linear polarization is represented by points in the equatorial plane, circular polarizations by the north pole and the south pole.

PROBLEMS

8.1 (a) Show that the absolute value of the correlation coefficient

$$j_{xy} = \frac{J_{xy}}{\sqrt{J_{xx}}\sqrt{J_{yy}}},$$

where J_{xx}, J_{yy} and J_{xy} are elements of the polarization matrix, does not exceed the value of the degree of polarization.

(b) Show also that the x- and the y-axes may always be so chosen that $|j_{xy}|$ is equal to the degree of polarization.

8.2 Show that the Stokes parameters of a quasi-monochromatic light beam satisfy the inequality

$$s_1^2 + s_2^2 + s_3^2 \leq s_0^2.$$

8.3 s_0, s_1, s_2 and s_3 are the Stokes parameters of a monochromatic light beam. Show that the expression

$$\frac{s_1^2 + s_2^2 + s_3^2}{s_0^2}$$

is invariant under rotation of the axes about the direction of propagation of the beam.

8.4 Derive an expression for the degree of polarization of a quasi-monochromatic light beam in terms of the Stokes parameters.

8.5 In the literature on scattering the following expression for the degree of polarization is often used:

$$\hat{P}(\mathbf{r}) = \frac{\left| I_x(\mathbf{r}) - I_y(\mathbf{r}) \right|}{I_x(\mathbf{r}) + I_y(\mathbf{r})}.$$

Here I_x and I_y denote the (averaged) intensities of the electric field in two mutually orthogonal directions perpendicular to the axis of the beam.

(a) Show that $\hat{P}$ is equal to the degree of polarization P, defined rigorously in Section 8.2.3, if and only if the polarization matrix $\mathbf{J}$ is diagonal.

(b) Determine the conditions under which the polarization matrix of a stochastic electromagnetic beam may be diagonalized by a rotation of the x- and y-axes about the direction of propagation of the beam.

(c) Assuming that the condition for diagonalizing the polarization matrix by rotation is satisfied, discuss the meaning of I_x and I_y in the above formula for $\hat{P}$.

8.6 Consider a monochromatic plane wave with Cartesian components $E_x \exp(-i\omega t)$, $E_y \exp(-i\omega t)$ of the complex electric field, in two mutually orthogonal directions perpendicular to the direction of propagation. Let

$$\mathbf{E} = \begin{bmatrix} E_x \\ E_y \end{bmatrix}.$$

E is known as the *Jones vector*.

Suppose that the wave passes through a linear non-image-forming device. The Jones vector of the emerging wave is then

$$\mathbf{E}' = \mathbf{LE},$$

where

$$\mathbf{L} = \begin{bmatrix} a & b \\ c & d \end{bmatrix}.$$

is known as the *instrument operator (matrix)*.

Determine the instrument operator of the device in each of the following cases.

(a) A compensator which introduces a phase difference $\delta = \varphi_y - \varphi_x$ between the phases φ_x and φ_y of the components E_x and E_y, respectively.

(b) An absorber which attenuates the x component of the electric field by the multiplicative factor $e^{-\alpha_x}$ and the y component of the electric field by the multiplicative factor $e^{-\alpha_y}$.

(c) A rotator which produces rotation of the electric field by an angle θ in the clockwise sense around the direction of propagation.

(d) A polarizer which transmits only the component of the electric field that makes an angle θ with the x direction, measured in the anticlockwise sense around the direction of propagation.

8.7 A quasi-monochromatic light beam, characterized by a polarization matrix $\mathbf{J}$ passes through a linear, non-image forming device.

(a) Show that the polarization matrix $\mathbf{J}'$ of the emergent beam is given by the formula

$$\mathbf{J}' = \mathbf{L}^* \mathbf{J} \mathbf{L}^T,$$

where $\mathbf{L}$ is the instrument matrix (see the previous problem) at the mean wavelength of the beam, the dagger denotes the Hermitian adjoint and T denotes the transpose.

(b) Show, with the help of the above transformation, that the averaged electric energy density of the beam does not change when the beam passes through a compensator or a rotator.

Derive also expressions for the averaged electric energy density of the beam after it has passed through (i) an absorber and (ii) a polarizer.

Unified theory of polarization
and coherence

So far we have treated the subjects of coherence and polarization independently of each other. Yet both are manifestations of the same physical phenomenon, namely of correlations between fluctuations in light beams. Coherence, we will soon see, arises from correlations between fluctuations at two or more points in space. Polarization, on the other hand, is a manifestation of correlation between fluctuating components of the electric field at a single point.

In recent times, more than 60 years since the publication of Zernike's basic paper on coherence and more than 160 years since Stokes introduced parameters which describe the state of polarization of a light beam, a unification of the phenomena of coherence and polarization was achieved.[1] In this chapter we will describe this development and we will give examples which demonstrate the usefulness of this more comprehensive formulation of correlation effects in stochastic electromagnetic beams.

9.1 The 2 × 2 cross-spectral density matrix of a
stochastic electromagnetic beam

The basic quantity of the unified theory of coherence and polarization of stochastic, statistically stationary, electromagnetic beams is the so-called *electric cross-spectral density matrix* $\mathbf{W}(\mathbf{r}_1, \mathbf{r}_2, \omega)$, which may be formally introduced as the Fourier transform of the *electric mutual coherence matrix*

$$
\mathbf{\Gamma}(\mathbf{r}_1, \mathbf{r}_2, \tau) = \begin{bmatrix} \langle E_x^*(\mathbf{r}_1, t)E_x(\mathbf{r}_2, t + \tau)\rangle & \langle E_x^*(\mathbf{r}_1, t)E_y(\mathbf{r}_2, t + \tau)\rangle \\ \langle E_y^*(\mathbf{r}_1, t)E_x(\mathbf{r}_2, t + \tau)\rangle & \langle E_y^*(\mathbf{r}_1, t)E_y(\mathbf{r}_2, t + \tau)\rangle \end{bmatrix}, \tag{1}
$$

where E_x and E_y are the components of the (complex) electric field vector, represented by analytic signals (see Section 2.3) in two mutually orthogonal directions perpendicular to

[1] E. Wolf, *Phys. Lett. A* **312** (2003), 263–267; *Opt. Lett.* **28** (2003), 1078–1080; and H. Roychowdhury and E. Wolf, *Opt. Commun.* **226** (2003), 57–60.

the axis of the beam (taken to be the z direction).[1] Explicitly one has

$$\mathbf{W}(\mathbf{r}_1, \mathbf{r}_2, \omega) \equiv \begin{bmatrix} W_{xx}(\mathbf{r}_1, \mathbf{r}_2, \omega) & W_{xy}(\mathbf{r}_1, \mathbf{r}_2, \omega) \\ W_{yx}(\mathbf{r}_1, \mathbf{r}_2, \omega) & W_{yy}(\mathbf{r}_1, \mathbf{r}_2, \omega) \end{bmatrix}$$

$$= \frac{1}{2\pi} \int_{-\infty}^{\infty} \mathbf{\Gamma}(\mathbf{r}_1, \mathbf{r}_2, \tau) e^{i\omega\tau} \, d\tau. \tag{2}$$

The matrices $\mathbf{\Gamma}$ and $\mathbf{W}$ are functions of two points, whereas the polarization matrix $\mathbf{J}$ of Section 8.1 is a function of just one point. This generalization is crucial for elucidating many polarization features of a fluctuating electromagnetic beam. In particular, as we will soon see, it makes it possible to determine how the degree of polarization may change as the beam propagates, whether in free space or in a medium.

We have introduced the cross-spectral density matrix $\mathbf{W}(\mathbf{r}_1, \mathbf{r}_2, \omega)$ as the Fourier transform of the mutual coherence matrix $\mathbf{\Gamma}(\mathbf{r}_1, \mathbf{r}_2, \tau)$ of the electric field. However, by analogy to the important result derived in Section 4.1 for stochastic scalar fields, the cross-spectral density matrix may also be expressed as a correlation matrix, i.e. in the form

$$\mathbf{W}(\mathbf{r}_1, \mathbf{r}_2, \omega) \equiv [W_{ij}(\mathbf{r}_1, \mathbf{r}_2, \omega)]$$

$$= \begin{bmatrix} \langle E_x^*(\mathbf{r}_1, \omega)E_x(\mathbf{r}_2, \omega) \rangle & \langle E_x^*(\mathbf{r}_1, \omega)E_y(\mathbf{r}_2, \omega) \rangle \\ \langle E_y^*(\mathbf{r}_1, \omega)E_x(\mathbf{r}_2, \omega) \rangle & \langle E_y^*(\mathbf{r}_1, \omega)E_y(\mathbf{r}_2, \omega) \rangle \end{bmatrix} \tag{3}$$

$$(i = x, y; \ j = x, y),$$

where $E_x(\mathbf{r}, \omega)$ and $E_y(\mathbf{r}, \omega)$ are members of suitably constructed statistical ensembles.[2]

This matrix is particularly useful in formulating the unified theory of coherence and polarization and for many applications.

9.2 The spectral interference law, the spectral degree of coherence and the spectral degree of polarization of stochastic electromagnetic beams

By analogy with our earlier discussions of fluctuating scalar wavefields we regard the state of coherence of an electromagnetic beam as the ability of the beam to produce fringes of

---

[1] A related, somewhat less general but nevertheless useful correlation matrix is the so-called beam coherence-polarization matrix, which was introduced by F. Gori, M. Santersiero, S. Vicalvi, R. Borghi and G. Guattari, *Pure Appl. Opt.* **7** (1988), 941–951.

Both matrices are restricted versions of the general 3 × 3 electric correlation matrix of the electric field (see M&W, Section 6.5.1).

It follows from the envelope representation of quasi-monochromatic signals (Section 2.3) that, when $|\tau| \ll 2\pi/\Delta\omega$ ($\Delta\omega$ being the effective bandwidth of the light), $\mathbf{\Gamma}(\mathbf{r}_1, \mathbf{r}_2, \tau)$ may be approximated by $\mathbf{\Gamma}(\mathbf{r}_1, \mathbf{r}_2, \tau) \approx \mathbf{\Gamma}(\mathbf{r}_1, \mathbf{r}, 0)\exp(-i\bar{\omega}\tau)$. This approximation corresponds to that given by Eq. (22) of Section 3.1 of the scalar theory.

[2] Proof of this result is given in Section 7 of J. Tervo, T. Setälä and A. T. Friberg, *J. Opt. Soc. Amer.* **A21** (2005), 2205–2215.

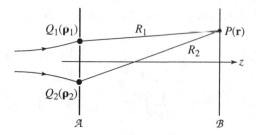

Fig. 9.1 Notation relating to Young's interference experiment with stochastic electromagnetic beams.

appropriate sharpness, in suitable interference experiments. Naturally, we associate a high degree of coherence with interference fringes of high visibility and a low degree of coherence with fringes of low visibility. The basic experiment which reveals the state of coherence is, of course, Young's interference experiment, that we discussed in Sections 3.1 and 4.2 in the context of the scalar theory. We will now discuss the experiment by taking into account the electromagnetic nature of the beams. Although the analysis will be similar to that which we presented for scalar fields, new questions now arise, such as the role which coherence and polarization play in propagation of partially coherent electromagnetic beams – a subject which is rather subtle and has been clarified only relatively recently.

Let us again consider a stochastic, statistically stationary, electromagnetic beam which propagates close to the z-axis and is incident on an opaque screen $\mathcal{A}$ in the plane $z = 0$, containing two small openings at points $Q_1(\boldsymbol{\rho}_1)$ and $Q_2(\boldsymbol{\rho}_2)$ (see Fig. 9.1).

We will determine the distribution of the averaged spectral energy density in a plane $\mathcal{B}$, placed at some distance beyond the plane $\mathcal{A}$ of the pinholes and parallel to it.

Let $\{\mathbf{E}(\mathbf{r}, \omega)\}$ represent the statistical ensemble of the electric vector at the point $P(\mathbf{r})$. A typical realization $\mathbf{E}(\mathbf{r}, \omega)$ of this ensemble is given, in terms of the realizations $\mathbf{E}(\boldsymbol{\rho}_1, \omega)$ and $\mathbf{E}(\boldsymbol{\rho}_2, \omega)$ of the electric field vector at the two pinholes, by the formula

$$\mathbf{E}(\mathbf{r}, \omega) = K_1\mathbf{E}(\boldsymbol{\rho}_1, \omega)e^{ikR_1} + K_2\mathbf{E}(\boldsymbol{\rho}_2, \omega)e^{ikR_2}, \tag{1}$$

where R_1 and R_2 are the distances from the points $Q(\boldsymbol{\rho}_1)$ and $Q(\boldsymbol{\rho}_2)$, respectively, to the point $P(\mathbf{r})$, and K_1 and K_2 are the same factors as in the scalar case [Eq. (3) of Section 3.1 with $\bar{\lambda}$ replaced by λ], again assuming that the angles of incidence and of diffraction at the pinholes are sufficiently small.

Let us now consider the spectral density $S(\mathbf{r}, \omega)$ of the field at the point $P(\mathbf{r})$. We may identify the spectral density (the spectrum) with the average electric energy density at that point. Apart from an unessential proportionality factor which depends on the choice of units, we have

$$S(\mathbf{r}, \omega) = \langle \mathbf{E}^*(\mathbf{r}, \omega) \cdot \mathbf{E}(\mathbf{r}, \omega) \rangle \tag{2a}$$

$$= \mathrm{Tr}\, \mathbf{W}(\mathbf{r}, \mathbf{r}, \omega), \tag{2b}$$

where Tr denotes the trace of the cross-spectral density matrix $\mathbf{W}(\mathbf{r}, \mathbf{r}, \omega)$ which we introduced in the previous section, evaluated for coincident points ($\mathbf{r}_1 = \mathbf{r}_2 \equiv \mathbf{r}$), i.e.

$$\text{Tr } \mathbf{W}(\mathbf{r}, \mathbf{r}, \omega) = \langle E_x^*(\mathbf{r}, \omega)E_x(\mathbf{r}, \omega)\rangle + \langle E_y^*(\mathbf{r}, \omega)E_y(\mathbf{r}, \omega)\rangle. \tag{3}$$

On substituting from Eq. (1) into Eq. (2a) we find that

$$\begin{aligned}
S(\mathbf{r}, \omega) = &\left|K_1\right|^2 S(\rho_1, \omega) + \left|K_2\right|^2 S(\rho_2, \omega) \\
&+ K_1^* K_2 \text{ Tr } \mathbf{W}(\rho_1, \rho_2, \omega)e^{ik(R_2 - R_1)} \\
&+ K_1 K_2^* \text{ Tr } \mathbf{W}(\rho_2, \rho_1, \omega)e^{-ik(R_2 - R_1)}.
\end{aligned} \tag{4}$$

Similarly to the scalar case, this formula may be rewritten in a physically more significant form by noting first that, if the pinhole at $Q_2(\rho_2)$ were closed, then $K_2 = 0$ and Eq. (4) would become $S(\mathbf{r}, \omega) = S^{(1)}(\mathbf{r}, \omega)$, where

$$S^{(1)}(\mathbf{r}, \omega) = \left|K_1\right|^2 S(\rho_1, \omega). \tag{5}$$

$S^{(1)}(\mathbf{r}, \omega)$ evidently represents the spectral density of the field reaching the point $P(\mathbf{r})$ through the pinhole $Q_1(\rho_1)$ only. A strictly similar expression is obtained for the spectral density, $S^{(2)}(\mathbf{r}, \omega)$, if the light reaches the point $P(\mathbf{r})$ only through the pinhole at $Q_2(\rho_2)$. Using these expressions and the relation

$$\text{Tr } \mathbf{W}(\rho_2, \rho_1, \omega) = [\text{Tr } \mathbf{W}(\rho_1, \rho_2, \omega)]^*, \tag{6}$$

which follows from the definition of the cross-spectral density matrix, we obtain from Eq. (4) the following expression for the spectral density at the point $P(\mathbf{r})$:

$$S(\mathbf{r}, \omega) = S^{(1)}(\mathbf{r}, \omega) + S^{(2)}(\mathbf{r}, \omega) + 2\sqrt{S^{(1)}(\mathbf{r}, \omega)}\sqrt{S^{(2)}(\mathbf{r}, \omega)} \; \mathcal{R}e[\eta(\rho_1, \rho_2, \omega)e^{ik(R_2 - R_1)}], \tag{7}$$

where $\mathcal{R}e$ denotes the real part and

$$\eta(\rho_1, \rho_2, \omega) = \frac{\text{Tr } \mathbf{W}(\rho_1, \rho_2, \omega)}{\sqrt{\text{Tr } \mathbf{W}(\rho_1, \rho_1, \omega)}\sqrt{\text{Tr } \mathbf{W}(\rho_2, \rho_2, \omega)}}, \tag{8a}$$

$$= \frac{\text{Tr } \mathbf{W}(\rho_1, \rho_2, \omega)}{\sqrt{S(\rho_1, \omega)}\sqrt{S(\rho_2, \omega)}}. \tag{8b}$$

If, as is usually the case, $S^{(2)}(\mathbf{r}, \omega) \approx S^{(1)}(\mathbf{r}, \omega)$, Eq. (7) reduces to

$$\begin{aligned}
S(\mathbf{r}, \omega) &= 2S^{(1)}(\mathbf{r}, \omega)\{1 + \mathcal{R}e[\eta(\rho_1, \rho_2, \omega)e^{i\delta}]\} \\
&= 2S^{(1)}(\mathbf{r}, \omega)\{1 + \left|\eta(\rho_1, \rho_2, \omega)\right|\cos[\alpha(\rho_1, \rho_2, \omega) + \delta]\},
\end{aligned} \tag{9}$$

where $\alpha(\rho_1, \rho_2, \omega)$ is the argument (phase) of η and

$$\delta = k(R_2 - R_1) \tag{10}$$

is the phase difference associated with propagation from $Q_1(\rho_1)$ to $P(\mathbf{r})$ and from $Q_2(\rho_2)$ to $P(\mathbf{r})$, respectively.

Equation (7) shows that the spectrum at a point $P(\mathbf{r})$ in the observation plane $\mathcal{B}$ is not just the sum of the spectra of the two beams reaching that point from the two pinholes but differs from it by the presence of the third term, which evidently represents the effect of interference. We may, therefore, refer to Eq. (7) as *the spectral interference law for the superposition of stochastic electromagnetic beams*. It is of the same form as the spectral interference law for scalar wavefields [Eq. (5) of Section 4.2], the only difference being that the factor $\eta(\rho_1, \rho_2, \omega)$ defined by Eq. (8a) now appears in place of the factor $\mu(\rho_1, \rho_2, \omega)$ defined by Eq. (6) of Section 4.2.

It is seen at once from the spectral interference law [Eq. (9)] that, as the path difference $R_2 - R_1$ and, consequently, the phase difference δ changes and the spectral density $S(\mathbf{r}, \omega)$ varies sinusoidally between the values

$$S_{max}(\mathbf{r}, \omega) = 2S^{(1)}(\mathbf{r}, \omega)\left\{1 + \left|\eta(\rho_1, \rho_2, \omega)\right|\right\} \tag{11a}$$

and

$$S_{min}(\mathbf{r}, \omega) = 2S^{(1)}(\mathbf{r}, \omega)\left\{1 - \left|\eta(\rho_1, \rho_2, \omega)\right|\right\}. \tag{11b}$$

Hence the *spectral visibility* $\mathcal{V}(\mathbf{r}, \omega)$ of the fringes is given by the expression

$$\mathcal{V}(\mathbf{r}, \omega) \equiv \frac{S_{max}(\mathbf{r}, \omega) - S_{min}(\mathbf{r}, \omega)}{S_{max}(\mathbf{r}, \omega) + S_{min}(\mathbf{r}, \omega)}$$

$$= \left|\eta(\rho_1, \rho_2, \omega)\right|. \tag{12}$$

Evidently

$$0 \le \left|\eta(\rho_1, \rho_2, \omega)\right| \le 1, \tag{13}$$

and the fringes are sharpest (visibility $\mathcal{V} = 1$) when $|\eta| = 1$ and there are no fringes at all ($\mathcal{V} = 0$) when $\eta = 0$. We may, therefore, identify $\eta(\rho_1, \rho_2, \omega)$, defined by Eq. (8a), with the (generally complex) *spectral degree of coherence* of the fluctuating electric field at the points $Q_1(\rho_1)$ and $Q_2(\rho_2)$. It can be determined experimentally from visibility measurements. The phase of η can also be determined from measurements, namely by determining the location of maxima in the interference pattern, in a manner similar to that in the scalar case [see the discussion which follows Eq. (13) of Section 4.2]. It is of interest to note that Eq. (8a), which defines the degree of coherence $\eta(\rho_1, \rho_2, \omega)$, depends only on the diagonal elements $W_{xx}(\rho_1, \rho_2, \omega)$ and $W_{yy}(\rho_1, \rho_2, \omega)$ of the correlation matrix $\mathbf{W}$. The physical reason for this fact can be understood by noting that two mutually orthogonal vector components $E_x\hat{\mathbf{x}}$ and $E_y\hat{\mathbf{y}}$ of the electric vector, with $\hat{\mathbf{x}}$ and $\hat{\mathbf{y}}$ denoting unit vectors along the x and y directions, do not interfere, because $\hat{\mathbf{x}} \cdot \hat{\mathbf{y}} = 0$. This is the essence of some of the classic Fresnel–Arago interference laws,[1] which were formulated before the electromagnetic theory of light was established. The fact that two mutually orthogonal components of the electric field do not interfere does not, of course, imply that they are necessarily uncorrelated.

Although the off-diagonal elements W_{xy} and W_{yx} do not contribute to the coherence properties of a beam, they play a role in specifying its polarization properties; in particular their spectral degree of polarization $\mathcal{P}(\mathbf{r}, \omega)$. This important quantity is defined by a formula strictly analogous to Eq. (26) of Section 8.2.3 with the polarization matrix $\mathbf{J}(\mathbf{r})$ replaced by the cross-spectral density matrix $\mathbf{W}(\mathbf{r}_1, \mathbf{r}_2, \omega)$, restricted to the situation where $\mathbf{r}_1 = \mathbf{r}_2 \equiv \mathbf{r}$, i.e.

$$\mathcal{P}(\mathbf{r}, \omega) = \sqrt{1 - \frac{4 \operatorname{Det} \mathbf{W}(\mathbf{r}, \mathbf{r}, \omega)}{[\operatorname{Tr} \mathbf{W}(\mathbf{r}, \mathbf{r}, \omega)]^2}}, \tag{14}$$

Det again denoting the determinant and Tr the trace, i.e.

$$\operatorname{Det} \mathbf{W}(\mathbf{r}, \mathbf{r}, \omega) = W_{xx}(\mathbf{r}, \mathbf{r}, \omega)W_{yy}(\mathbf{r}, \mathbf{r}, \omega) - W_{xy}(\mathbf{r}, \mathbf{r}, \omega)W_{yx}(\mathbf{r}, \mathbf{r}, \omega), \tag{15a}$$

$$\operatorname{Tr} \mathbf{W}(\mathbf{r}, \mathbf{r}, \omega) = W_{xx}(\mathbf{r}, \mathbf{r}, \omega) + W_{yy}(\mathbf{r}, \mathbf{r}, \omega). \tag{15b}$$

Since the degree of polarization is expressed in terms of both the trace and the determinant of the correlation matrix $\mathbf{W}$, it does indeed depend, in general, not only on the diagonal but also on the off-diagonal elements of the matrix. Finally we stress that, whilst the spectral degree of coherence depends on the behavior of the electric field at two points, the spectral degree of polarization depends on the behavior of the electric field at a single point only.

9.3 Determination of the cross-spectral density matrix from experiments

We will now show how the elements of the cross-spectral density matrix of the electric field of a stochastic beam may be determined experimentally.

Let us suppose that we again perform Young's interference experiment, with incident light that is filtered so that it becomes effectively quasi-monochromatic around the

---

[1] For accounts of the Fresnel–Arago interference laws see, for example, E. Collett, *Am. J. Phys.* **39** (1971), 1483–1495; E. Collett, *Polarized Light Fundamentals and Applications* (Marcel Dekker, New York, 1993), Chapter 12; or C. Brosseau, *Fundamentals of Polarized Light* (Wiley, New York, 1998), p. 6.

A derivation of the Fresnel–Arago interference laws which incorporates effects of coherence is given in M. Mujat, A. Dogariu and E. Wolf, *J. Opt. Soc. Amer.* **21** (2004), 2414–2417.

frequency ω. According to the spectral interference law given by Eq. (7) of Section 9.2, the spectral density at the point $P(\mathbf{r})$ may be expressed in the form

$$S(\mathbf{r}, \omega) = S^{(1)}(\mathbf{r}, \omega) + S^{(2)}(\mathbf{r}, \omega)$$
$$+ 2\sqrt{S^{(1)}(\mathbf{r}, \omega)}\sqrt{S^{(2)}(\mathbf{r}, \omega)}\,|\eta(\rho_1, \rho_2, \omega)|\cos\left[\alpha(\rho_1, \rho_2, \omega) + \delta\right], \tag{1}$$

where δ is the phase difference $k(R_2 - R_1)$. In this formula $S^{(1)}(\mathbf{r})$ is the spectral density of the field which would be observed at $P(\mathbf{r})$ with only the pinhole at $Q_1(\rho_1)$ opened, $S^{(2)}(\mathbf{r})$ having a similar meaning. Further,

$$\eta(\rho_1, \rho_2, \omega) \equiv |\eta(\rho_1, \rho_2, \omega)| e^{i\alpha(\rho_1, \rho_2, \omega)}$$
$$= \frac{\mathrm{Tr}\, \mathbf{W}(\rho_1, \rho_2, \omega)}{\sqrt{S(\rho_1, \omega)}\sqrt{S(\rho_2, \omega)}} \tag{2}$$

is the spectral degree of coherence of the electric field at the pinholes [Eq. (8b) of Section 9.2].

Suppose that polarizers Π_1 and Π_2, which transmit only the x components of the incident electric field, are placed in front of the pinholes. The cross-spectral density matrix, $[\mathbf{W}\rho_1, \rho_2, \omega)]^+$ say, of the light emerging from the pinholes is

$$[\mathbf{W}(\rho_1, \rho_2, \omega)]^+ = \begin{bmatrix} \langle E_x^*(\rho_1, \omega)E_x(\rho_2, \omega)\rangle & 0 \\ 0 & 0 \end{bmatrix}. \tag{3}$$

Clearly

$$[\mathrm{Tr}\, \mathbf{W}(\rho_1, \rho_2, \omega)]^+ = W_{xx}(\rho_1, \rho_2, \omega), \tag{4}$$

i.e. the trace of the cross-spectral density matrix of the transmitted light is the leading-diagonal element of the cross-spectral density matrix $\mathbf{W}$ of the light incident on the pinholes. Using Eq. (4), Eq. (2) gives

$$W_{xx}(\rho_1, \rho_2, \omega) = \sqrt{S_x(\rho_1, \omega)}\sqrt{S_x(\rho_2, \omega)}\,\eta_{xx}(\rho_1, \rho_2, \omega), \tag{5}$$

where the subscripts on the quantities on the right indicate that the values of the spectra and of the spectral degree of coherence pertain to this experimental set-up, i.e. when only the x component is transmitted.

In a similar way, if polarizers which transmit only the y component of the incident beam are placed in front of the pinholes, one obtains in place of Eq. (5) the analogous formula

$$W_{yy}(\rho_1, \rho_2, \omega) = \sqrt{S_y(\rho_1, \omega)}\sqrt{S_y(\rho_2, \omega)}\,\eta_{yy}(\rho_1, \rho_2, \omega). \tag{6}$$

An expression for the off-diagonal elements of the matrix $\mathbf{W}(\rho_1, \rho_2, \omega)$ may be obtained as follows: linear polarizers are again placed in front of the pinholes. The polarizer at $Q_1(\rho_1)$ transmits the x component and the polarizer at $Q_2(\rho_2)$ transmits the y component of the incident light. In addition a rotator is placed behind the polarizer at the pinhole $Q_2(\rho_2)$, which rotates the transmitted field component through 90° in the clockwise sense about the beam axis [the (positive) z direction]. The cross-spectral density matrix of the light emerging from the pinholes now is

$$[\mathbf{W}(\rho_1, \rho_2, \omega)]^+ = \begin{bmatrix} \langle E_x^*(\rho_1, \omega)E_y(\rho_2, \omega)\rangle & 0 \\ 0 & 0 \end{bmatrix}. \tag{7}$$

Hence, with this arrangement,

$$[\mathrm{Tr}\, \mathbf{W}(\rho_1, \rho_2, \omega)]^+ = W_{xy}(\rho_1, \rho_2, \omega). \tag{8}$$

Equation (2) now gives, in obvious notation,

$$W_{xy}(\rho_1, \rho_2, \omega) = \sqrt{S_x(\rho_1, \omega)}\sqrt{S_y(\rho_2, \omega)}\eta_{xy}(\rho_1, \rho_2, \omega). \tag{9}$$

With a strictly analogous arrangement one can obtain the other off-diagonal element of the $\mathbf{W}$ matrix in the form

$$W_{yx}(\rho_1, \rho_2, \omega) = \sqrt{S_y(\rho_1, \omega)}\sqrt{S_x(\rho_2, \omega)}\,\eta_{yx}(\rho_1, \rho_2, \omega). \tag{10}$$

The spectra $S_x(\rho_1, \omega)$ and $S_y(\rho_2, \omega)$ may be measured with the help of the usual spectroscopic devices and, as we noted earlier (see the discussion following Eq. (13) in Section 9.2), the spectral degree of coherence $\eta_{ij}(\rho_1, \rho_2, \omega)$, $(i, j = x, y)$, may be determined from interference experiments. Thus, we have shown how all four elements of the 2×2 cross-spectral density matrix $\mathbf{W}(\rho_1, \rho_2, \omega)$ may be measured.

9.4 Changes in random electromagnetic beams on propagation

We showed in Section 4.2 that the spectrum of light may change on propagation. Such changes may be said to be induced by the coherence properties of the source. The analysis in Section 4.2 was based on scalar theory. One might expect that spectral changes will also be generated when the electromagnetic nature of the light is taken into account. One might also expect that there will be other changes on propagation, for example, in the degree of polarization of the beam and in its state of polarization, i.e. changes in the size, the shape and the orientation of the polarization ellipse of the polarized portion of the beam. In this section we will consider such changes which may also be said to be *correlation-induced*. A treatment using the *cross-spectral density matrix*, which we will sometimes just call the *correlation matrix* for short, provides a unified approach to determining changes of this kind. To elucidate them we will first determine how the matrix changes as the beam propagates.

9.4.1 Propagation of the cross-spectral density matrix of a stochastic electromagnetic beam – general formulas

Let $\{\mathbf{E}^{(0)}(\rho', \omega)\}$ and $\{\mathbf{E}(\mathbf{r}, \omega)\}$ be the statistical ensembles, introduced in Section 4.1, of the spectral components at frequency ω of the fluctuating electric field vectors at a point $Q(\rho')$

in the source plane $z = 0$ and at the point $P(\mathbf{r}) \equiv P(\rho, z)$ in the half-space $z > 0$, respectively (see Fig. 9.2). Then [see M&W, Eqs. (5.6–14) and (5.6–17)]

$$E_j(\mathbf{r}, \omega) = e^{ikz} \int_{(z=0)} E_j^{(0)}(\rho', \omega) G(\rho - \rho', z; \omega) d^2\rho', \tag{1}$$

where $j = x$ or y represents the components of the electric field vector in two mutually orthogonal directions, perpendicular to the axis of the beam (the z direction). Further,

$$G(\rho - \rho', z; \omega) = -\frac{ik}{2\pi z} \exp\left(\frac{ik(\rho - \rho')^2}{2z}\right) \tag{2}$$

($k = \omega/c$, c being the speed of light in vacuum) is Green's function for paraxial propagation from the source point $Q(\rho')$ to the field point $P(\mathbf{r} \equiv \rho, z)$.

On substituting for E_j from Eq. (1) into Eq. (3) of Section 9.1 one obtains the following expression for the spectral correlation matrix $\mathbf{W}$ at a pair of points located in any transverse plane $z = \text{constant} > 0$ in terms of the 2×2 correlation matrix $\mathbf{W}^{(0)}$ of the electric field vector at pairs of points in the source plane $z = 0$:

$$\mathbf{W}(\rho_1, \rho_2, z; \omega) \equiv \iint_{z=0} \mathbf{W}^{(0)}(\rho_1', \rho_2', \omega) K(\rho_1 - \rho_1', \rho_2 - \rho_2', z; \omega) d^2\rho_1' \, d^2\rho_2', \tag{3}$$

where

$$K(\rho_1 - \rho_1', \rho_2 - \rho_2', z; \omega) = G^*(\rho_1 - \rho_1', z; \omega) G(\rho_2 - \rho_2', z; \omega). \tag{4}$$

Equation (3) applies to propagation in free space. It is not difficult to generalize it to propagation in any linear medium whether deterministic or random.[1]

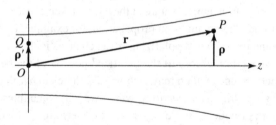

Fig. 9.2 Notation relating to the propagation of an electromagnetic beam.

---

[1] Some applications of the generalized formulas have been used in studies of propagation of random electromagnetic beams in the turbulent atmosphere. See, for example, H. Roychowdhury, S. A. Ponomarenko and E. Wolf, *J. Mod. Opt.* **52** (2005), 1611–1618; M. Salem, O. Korotkova, A. Dogariu and E. Wolf, *Waves in Random Media* **14** (2004), 513–523; and O. Korotkova, M. Salem, A. Dogariu and E. Wolf, *Waves in Random and Complex Media* **15** (2005), 353–364.

9.4.2 Propagation of the cross-spectral density matrix of an electromagnetic Gaussian Schell-model beam

We will illustrate the usefulness of the preceding analysis by a number of examples. We will restrict ourselves to correlation-induced changes in so-called *electromagnetic Gaussian Schell-model beams*. These are model beams which are generalizations of the scalar Gaussian Schell-model beams that we encountered in Section 5.3. The elements of the correlation matrix of planar sources which generate such beams are

$$W_{ij}^{(0)}(\boldsymbol{\rho}_1', \boldsymbol{\rho}_2', \omega) = \sqrt{S_i^{(0)}(\boldsymbol{\rho}_1', \omega)} \sqrt{S_j^{(0)}(\boldsymbol{\rho}_2', \omega)} \, \mu_{ij}^{(0)}(\boldsymbol{\rho}_2' - \boldsymbol{\rho}_1', \omega),$$

$$(i = x, y; \, j = x, y). \tag{5}$$

Here $\boldsymbol{\rho}_1'$ and $\boldsymbol{\rho}_2'$ are two-dimensional position vectors of points in the source plane $z = 0$, $S_i^{(0)}(\boldsymbol{\rho}_1', \omega)$ and $S_j^{(0)}(\boldsymbol{\rho}_2', \omega)$ denote the spectral densities of the i and the j components, respectively, of the electric field vector in the source plane $z = 0$, and $\mu_{ij}^{(0)}(\boldsymbol{\rho}_2' - \boldsymbol{\rho}_1', \omega)$ denotes the degree of correlation between the components E_i at $\boldsymbol{\rho}_1'$ and E_j at $\boldsymbol{\rho}_2'$. Both these quantities are given by Gaussian functions, viz.,

$$S_i^{(0)}(\boldsymbol{\rho}', \omega) = A_i^2 \exp[-\rho'^2/(2\sigma_i^2)], \tag{6a}$$

$$\mu_{ij}^{(0)}(\boldsymbol{\rho}_2' - \boldsymbol{\rho}_1', \omega) = B_{ij} \exp[-(\boldsymbol{\rho}_2' - \boldsymbol{\rho}_1')^2/(2\delta_{ij}^2)], \quad (i = x, y; \, j = x, y). \tag{6b}$$

The parameters A_i, B_{ij}, σ_i and δ_{ij} are independent of position but may depend on the frequency ω. However, they cannot be chosen quite arbitrarily. In particular

$$B_{ij} = 1 \quad \text{when } i = j, \tag{7a}$$

$$|B_{ij}| \leq 1 \quad \text{when } j \neq i, \tag{7b}$$

and

$$B_{ij} = B_{ji}^*, \tag{7c}$$

where the asterisk denotes the complex conjugate. Further,

$$\delta_{ji} = \delta_{ij}. \tag{7d}$$

The constraint (7a) follows at once from the fact that, when $j = i$, $\mu_{ij}^{(0)}(\boldsymbol{\rho}_2' - \boldsymbol{\rho}_1', \omega)$ is just the usual correlation coefficient which necessarily has the value unity when $\boldsymbol{\rho}_2' - \boldsymbol{\rho}_1' = 0$. The inequality (7b) follows from the fact that the correlation coefficient μ_{ij} cannot exceed unity in absolute value.[1] The relations (7c) and (7d) follows from the fact that

---

[1] See Appendix A in O. Korotkova, M. Salem and E. Wolf, *Opt. Commun.* **233** (2004), 225–230.

$W_{ij}(\mathbf{r}_1, \mathbf{r}_2, \omega) = W_{ji}^*(\mathbf{r}_2, \mathbf{r}_1, \omega)$, which is an immediate consequence of the definition of the correlation matrix $\mathbf{W}$. The variances δ_{ij}^2 and the coefficients B_{ij} have to satisfy some additional constraints, which are consequences of the fact that the correlation matrix is necessarily non-negative definite. Such constraints have been discussed in several papers.[1] In addition there are also constraints that have to be satisfied in order that the source generates a beam. If

$$\sigma_x = \sigma_y \equiv \sigma \tag{8}$$

the constraints are[2]

$$\frac{1}{4\sigma^2} + \frac{1}{\delta_{ij}^2} \ll \frac{2\pi^2}{\lambda^2}, \qquad (i = x, y; \; j = x, y). \tag{9}$$

It is shown in Appendix III that when the condition (8) is satisfied, i.e. when the r.m.s. widths of the spectral densities $S_x^{(0)}(\rho', \omega)$ and $S_y^{(0)}(\rho', \omega)$ of the x and y components of the electric field at each point in the source plane $z = 0$ are equal, the spectral degree of polarization $\mathcal{P}^{(0)}(\rho, \omega)$ is independent of ρ, i.e. it has the same value at every point of the source plane.

Let us now consider an electromagnetic Gaussian Schell-model beam, i.e. a beam generated by a Gaussian Schell-model source.[3] Its cross-spectral density matrix $\mathbf{W}(\rho_1, \rho_2, z; \omega)$ may be calculated by substituting from Eqs. (5) and (6) into the propagation law [Eq. (3)], with Green's function $G(\rho - \rho', z; \omega)$ in the propagation kernel $K(\rho_1 - \rho_1', \rho_2 - \rho_2'; z, \omega)$, defined by Eq. (4), being given by the paraxial approximation (2). One finds after a straightforward but long calculation that, assuming that the constraint given in Eq. (9) applies,

$$W_{ij}(\rho_1, \rho_2; z, \omega) = \frac{A_i A_j B_{ij}}{\Delta_{ij}^2(z)} \exp\left[-\frac{(\rho_1 + \rho_2)^2}{8\sigma^2 \Delta_{ij}^2(z)}\right] \exp\left[-\frac{(\rho_2 - \rho_1)^2}{2\Omega_{ij}^2 \Delta_{ij}^2(z)}\right] \exp\left[-\frac{ik(\rho_2^2 - \rho_1^2)}{2R_{ij}(z)}\right], \tag{10}$$

where the quantities $\Delta_{ij}^2(z)$, sometimes called the *beam-expansion coefficients*, are given by the expressions

$$\Delta_{ij}^2(z) = 1 + \left(\frac{z}{k\sigma\Omega_{ij}}\right)^2, \qquad \frac{1}{\Omega_{ij}^2} = \frac{1}{4\sigma^2} + \frac{1}{\delta_{ij}^2} \tag{11a}$$

[1] See F. Gori, M. Santarsiero, G. Piquero, R. Borghi, A. Mondello and R. Simon, *J. Pure Appl. Opt.* **3** (2001), 1–9 and H. Roychowdhury and O. Korotkova, *Opt. Commun.* **249** (2005), 379–385 which provides a somewhat more general treatment.
[2] O. Korotkova, M. Salem and E. Wolf, *Opt. Lett.* **29** (2004), 1173–1175.
[3] Methods for generating Gaussian Schell-model sources and Gaussian Schell-model beams have been described by G. Piquero, F. Gori, P. Roumanini, M. Santarsiero, R. Borghi and A. Mondello in *Opt. Commun.* **208** (2002), 9–16 and T. Shirai, O. Korotkova and E. Wolf, *J. Opt. A: Pure Appl. Opt.* **7** (2005), 232–237.

and

$$R_{ij}(z) = \left[1 + \left(\frac{k\sigma\Omega_{ij}}{z}\right)^2\right] z. \tag{11b}$$

Of special interest is the case in which the field points are in the far zone. One then finds that, provided that Eq. (10) holds, the elements of the cross-spectral density matrix of the far field (denoted by superscript ∞) are given by the formula[1]

$$W_{ij}^{(\infty)}(r_1\mathbf{s}_1, r_2\mathbf{s}_2, \omega) = k^2 \cos\theta_1 \cos\theta_2 \frac{A_i A_j B_{ij}}{(a_{ij}^2 - b_{ij}^2)} \exp\{-2k^2[\alpha_{ij} - \beta_{ij}(\mathbf{s}_1 \cdot \mathbf{s}_2)]\}$$

$$\times \frac{\exp[ik(r_2 - r_1)]}{r_1 r_2}, \qquad (i = x, y; \ j = x, y), \tag{12}$$

as $kr_1 \to \infty$ and $kr_2 \to \infty$, with the directions, specified by the unit vectors $\mathbf{s}_1$ and $\mathbf{s}_2$, being kept fixed (see Fig. 9.3) and with

$$\alpha_{ij} = \frac{a_{ij}}{4(a_{ij}^2 - b_{ij}^2)}, \qquad \beta_{ij} = \frac{b_{ij}}{4(a_{ij}^2 - b_{ij}^2)}, \tag{13a}$$

where

$$a_{ij} = \frac{1}{2}\left[\frac{1}{2\sigma^2} + \frac{1}{\delta_{ij}^2}\right], \qquad b_{ij} = \frac{1}{2\delta_{ij}^2}. \tag{13b}$$

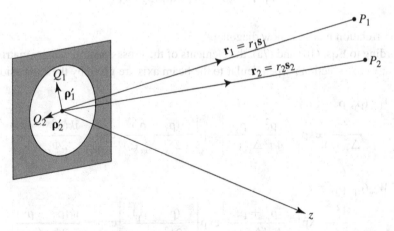

Fig. 9.3 Notation relating to the arguments of the correlation matrix at points in the far field of an electromagnetic Gaussian Schell-model source.

[1] O. Korotkova, M. Salem and E. Wolf, *Opt. Lett.* **29** (2004), 1173–1175.

9.4.3 Examples of correlation-induced changes in stochastic electromagnetic beams on propagation

Equation (10) gives the elements of the cross-spectral density matrix at pairs of points in any cross-section $z = \text{constant} > 0$ of electromagnetic Gaussian Schell-model beams of a wide class, namely those for which the constraint given by Eq. (8), i.e. $\sigma_x = \sigma_y = \sigma$, holds. As we have already noted, the constraint ensures that the degree of polarization is the same at every point in the source plane $z = 0$.

On substituting for the matrix elements given by Eq. (10) in Eqs. (2) and (14) of Section 9.2 one can determine the changes in the spectral density $S(\mathbf{r}, \omega)$ and in the spectral degree of polarization $\mathcal{P}(\mathbf{r}, \omega)$ throughout the half-space $z > 0$ into which the beam propagates. Using Eq. (10) one may also determine changes in the complete state of polarization of the beam at each point in that half-space, i.e. the shape and the orientation of the polarization ellipse of the polarized portion of the beam. We will discuss such changes shortly.

For simplicity we will assume that the x and y components of the electric field in the source plane are uncorrelated, i.e. that

$$\mu_{xy}^{(0)}(\boldsymbol{\rho}_2' - \boldsymbol{\rho}_1', \omega) = \mu_{yx}^{(0)}(\boldsymbol{\rho}_2' - \boldsymbol{\rho}_1', \omega) = 0. \tag{14}$$

According to Eq. (6b) this implies that

$$B_{xy} = B_{yx} = 0, \tag{15}$$

and it is clear from Eq. (5) that, in this case,

$$W_{xy}^{(0)}(\boldsymbol{\rho}_1', \boldsymbol{\rho}_2'; \omega) = W_{xy}^{(0)}(\boldsymbol{\rho}_1', \boldsymbol{\rho}_2'; \omega) = 0, \tag{16}$$

i.e. the correlation matrix is now diagonal.

According to Eqs. (10) and (7a), the elements of the cross-spectral density matrix of the field in any cross-section perpendicular to the beam axis are given by the expressions

$$
\begin{aligned}
&W_{xx}(\boldsymbol{\rho}_1, \boldsymbol{\rho}_2, z; \omega) \\
&= \frac{A_x^2}{\Delta_{xx}^2(z)} \exp\left[-\frac{\rho_1^2 + \rho_2^2}{4\sigma^2 \Delta_{xx}^2(z)}\right] \exp\left[-\frac{(\boldsymbol{\rho}_2 - \boldsymbol{\rho}_1)^2}{2\delta_{xx}^2 \Delta_{xx}^2(z)}\right] \exp\left[-\frac{ik(\rho_2^2 - \rho_1^2)}{2\Phi_{xx}(z)}\right],
\end{aligned} \tag{17a}
$$

$$
\begin{aligned}
&W_{yy}(\boldsymbol{\rho}_1, \boldsymbol{\rho}_2, z; \omega) \\
&= \frac{A_y^2}{\Delta_{yy}^2(z)} \exp\left[-\frac{\rho_1^2 + \rho_2^2}{4\sigma^2 \Delta_{yy}^2(z)}\right] \exp\left[-\frac{(\boldsymbol{\rho}_2 - \boldsymbol{\rho}_1)^2}{2\delta_{yy}^2 \Delta_{yy}^2(z)}\right] \exp\left[-\frac{ik(\rho_2^2 - \rho_1^2)}{2\Phi_{yy}(z)}\right],
\end{aligned} \tag{17b}
$$

$$W_{xy}(\boldsymbol{\rho}_1, \boldsymbol{\rho}_2, z; \omega) = W_{yx}(\boldsymbol{\rho}_1, \boldsymbol{\rho}_2, z; \omega) = 0. \tag{17c}$$

In the formulas (17a) and (17b) the beam expansion coefficients are given by the formulas

$$\Delta_{xx}^2(z) = 1 + \frac{1}{(k\sigma)^2}\left(\frac{1}{4\sigma^2} + \frac{1}{\delta_{xx}^2}\right)z^2, \tag{18a}$$

$$\Delta_{yy}^2(z) = 1 + \frac{1}{(k\sigma)^2}\left(\frac{1}{4\sigma^2} + \frac{1}{\delta_{yy}^2}\right)z^2, \tag{18b}$$

and, according to Eq. (11b),

$$\Phi_{xx}(z) = \left(1 + \frac{1}{\Delta_{xx}^2(z)}\right)z, \tag{19a}$$

$$\Phi_{yy}(z) = \left(1 + \frac{1}{\Delta_{yy}^2(z)}\right)z. \tag{19b}$$

Let us now make use of these expressions to elucidate, by means of a few examples, correlation-induced changes in an electromagnetic Gaussian Schell-model beam on propagation.

We have from Eqs (2b) and (17) that

$$S(\mathbf{r}, \omega) = \mathrm{Tr}\,\mathbf{W}(\boldsymbol{\rho}, \boldsymbol{\rho}, z;\ \omega) \equiv W_{xx}(\boldsymbol{\rho}, \boldsymbol{\rho}, z;\ \omega) + W_{yy}(\boldsymbol{\rho}, \boldsymbol{\rho}, z;\ \omega)$$

$$= \frac{A_x^2}{\Delta_{xx}^2(z)}\exp\left[-\frac{\rho^2}{2\sigma^2\Delta_{xx}^2(z)}\right] + \frac{A_y^2}{\Delta_{yy}^2(z)}\exp\left[-\frac{\rho^2}{2\sigma^2\Delta_{yy}^2(z)}\right]. \tag{20}$$

Since in the present case the correlation matrix is diagonal, we obtain, on using Eqs. (17), the following expressions for the determinant:

$$\mathrm{Det}\,\mathbf{W}(\boldsymbol{\rho}, \boldsymbol{\rho}, z;\ \omega) = W_{xx}(\boldsymbol{\rho}, \boldsymbol{\rho}, z;\ \omega)\,W_{yy}(\boldsymbol{\rho}, \boldsymbol{\rho}, z;\ \omega)$$

$$= \frac{A_x^2 A_y^2}{\Delta_{xx}^2(z)\,\Delta_{yy}^2(z)}\exp\left[-\left(\frac{1}{\Delta_{xx}^2(z)} + \frac{1}{\Delta_{yy}^2(z)}\right)\frac{\rho^2}{2\sigma^2}\right]. \tag{21}$$

On substituting from Eqs. (20) and (21) into Eq. (14) of Section 9.2, we obtain an expression for the degree of polarization at any point in the beam. A few examples of such changes in planes $z = $ constant > 0 are also shown in Fig. 9.4. The behavior of the spectral degree of polarization along the axis is shown in Fig. 9.5. The considerable differences between these figures reveal that the changes in the degree of polarization on propagation depend very sensitively on the values of the parameters that specify the source.

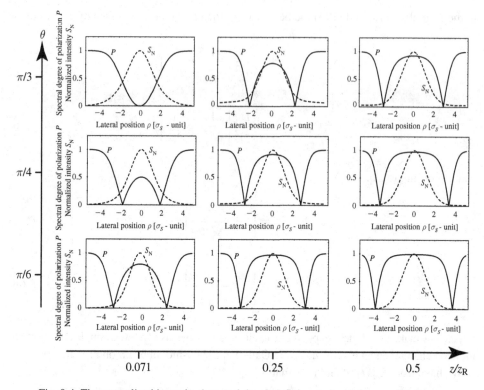

Fig. 9.4 The normalized intensity (spectral density) $S_N(\rho, z; \omega)$ and the spectral degree of polarization $\mathcal{P}(\rho, z; \omega)$ of a stochastic electromagnetic beam propagating into the half-space $z > 0$. The field in the source plane $z = 0$ is represented by a cross-spectral density matrix with elements

$$W_{xx}(\rho_1', \rho_2', \omega) = F(\rho_1', \rho_2', \omega)\cos^2\theta,$$

$$W_{yy}(\rho_1', \rho_2', \omega) = F(\rho_1', \rho_2', \omega)\exp\left(-\frac{(\rho_2' - \rho_1')^2}{2\delta_{yy}^2}\right)\sin^2\theta,$$

$$W_{xy}(\rho_1', \rho_2', \omega) = W_{yx}(\rho_1', \rho_2', \omega) = 0,$$

with

$$F(\rho_1', \rho_2', \omega) = S_0(\omega)\exp\left(-\frac{\rho_1'^2 + \rho_2'^2}{4\sigma_s^2}\right).$$

The source generates a linearly polarized, spatially coherent Gaussian electromagnetic beam, with polarization angle θ and with the effective width σ_s, incident on a random phase screen. The parameter δ_{yy} is a constant relating to the properties of the screen. The values of the parameters were taken as $\sigma_s = 1$ mm, $\delta_{yy} = 0.1$ mm and $\lambda = 632.8$ nm. The distance of propagation z is normalized by the Rayleigh range $z_R = 2k\sigma_s^2$ ($k = \omega/c$). [Reproduced from T. Shirai and E. Wolf, *J. Opt. Soc. Amer.* **A21** (2004), 1907–1916.]

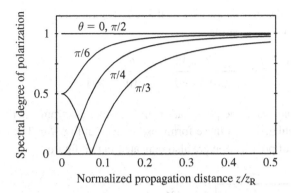

Fig. 9.5 The spectral degree of polarization $\mathcal{P}(0, z, \omega)$ of a stochastic electromagnetic beam, with the same parameters as in Fig. 9.4, along the axis of the beam (the z-axis), for some selected values of the polarization angle θ. The propagation distance is normalized by the Rayleigh range $z_R = 2k\sigma_s^2$ of the incident beam. [Reproduced from T. Shirai and E. Wolf, *J. Opt. Soc. Amer.* **A21** (2004), 1907–1916.]

It may seem surprising that, in general, the degree of polarization changes as the beam propagates. The reason for such changes is the difference between the correlation coefficients which specify the correlation between the x and the y components of the electric field vector at the two source points. Unless the correlation coefficients are the same, the expansion coefficients $\Delta_{xx}^2(z)$ and $\Delta_{yy}^2(z)$, defined by Eqs. (18), have different values. Consequently the diagonal elements W_{xx} and W_{yy} of the correlation matrix, given by Eqs. (17), expand at different rates. This leads to different behaviors of the two matrix elements, resulting in changes in the degree of polarization as the beam propagates.

One might expect that not only does the degree of polarization, in general, change on propagation but also that the shape and the orientation of the polarization ellipse of the polarized portion of the beam change. This indeed is the case.[1] Because the derivation of the pertinent formulas is rather lengthy, we will only state them here.

One finds that, in terms of the elements of the correlation matrix, the squares of the major and the minor semi-axes of the polarization ellipse are given by the formulas

$$a^2(\rho, z; \omega) = \frac{1}{8}\left[\sqrt{(W_{xx} - W_{yy})^2 + 4\left|W_{xy}\right|^2} + \sqrt{(W_{xx} - W_{yy})^2 + 4[\mathcal{R}e\, W_{xy}]^2}\right], \quad (22a)$$

$$b^2(\rho, z; \omega) = \frac{1}{8}\left[\sqrt{(W_{xx} - W_{yy})^2 + 4\left|W_{xy}\right|^2} - \sqrt{(W_{xx} - W_{yy})^2 + 4[\mathcal{R}e\, W_{xy}]^2}\right], \quad (22b)$$

where $\mathcal{R}e$ denotes the real parts.

---

[1] See O. Korotkova and E. Wolf, *Opt. Commun.* **246** (2005), 35–43.

The angle θ_0 which the major axis of the polarization ellipse makes with the x-axis is found to be given by the formula

$$\theta_0(\rho, z; \omega) = \frac{1}{2} \arctan \left[\frac{2 \, \mathcal{R}e \, W_{xy}(\rho, z; \omega)}{W_{xx}(\rho, z; \omega) - W_{yy}(\rho, z; \omega)} \right] \quad (-\pi/2 < \theta_0 \leq \pi/2). \quad (23)$$

An example of changes in the polarization ellipse of an electromagnetic Gaussian Schell-model beam, calculated from these formulas, is given in Fig. 9.6. The corresponding values of the degree of polarization are also indicated in that figure.

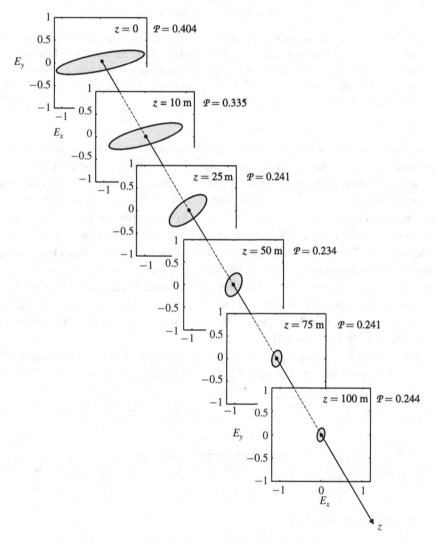

Fig. 9.6 Changes in the polarization ellipse and the degree of polarization associated with a typical Gaussian Schell-model beam on propagation in free space. [Reproduced from O. Korotkova and E. Wolf, *Opt. Commun.* **246** (2005), 35–43.]

9.4.4 Coherence-induced changes of the degree of polarization in Young's interference experiment[1]

In the previous section we studied correlation-induced changes in the polarization properties of a Gaussian Schell-model beam on propagation in free space. Somewhat similar changes may take place when two correlated beams are superposed, for example, in a Young interference experiment. In this section we will briefly discuss the change in the degree of polarization in Young's interference pattern as the degree of coherence of the light incident on the pinholes is changed.

Consider a stochastic, statistically stationary electromagnetic beam propagating close to the z-axis and suppose that an opaque screen $\mathcal{A}$ is placed across the plane $z = 0$ with small openings at points $Q_1(\rho_1)$ and $Q_2(\rho_2)$ and that measurements are made at a point $P(\mathbf{r})$ in the observation plane $\mathcal{B}$, which is parallel to the screen $\mathcal{A}$ in the half-space $z > 0$ (see Fig. 9.1).

Let $\{\mathbf{E}(\rho_1, \omega)\}$ and $\{\mathbf{E}(\rho_2, \omega)\}$ be the statistical ensembles of the electric field vectors at the two pinholes. If $\{\mathbf{E}(\mathbf{r}, \omega)\}$ denotes the statistical ensemble of the electric field vector at the point $P(\mathbf{r})$ then we have for each member of this ensemble, as in Eq. (1) of Section 9.2,

$$\mathbf{E}(\mathbf{r}, \omega) = K_1 \mathbf{E}(\rho_1, \omega)e^{ikR_1} + K_2 \mathbf{E}(\rho_2, \omega)e^{ikR_2}, \tag{24}$$

where K_1 and K_2 are the same geometrical factors as we encountered before [Eq. (3) of Section 3.1 (with $\bar{\lambda}$ now replaced by λ)]. We again assume that the angles of incidence and of diffraction at each pinhole are small.

To determine the degree of polarization at a point $P(\mathbf{r})$ in the observation plane $\mathcal{B}$ we must first determine the elements of the correlation matrix

$$\mathbf{W}(\mathbf{r}, \mathbf{r}, \omega) = [W_{ij}(\mathbf{r}, \mathbf{r}, \omega)] = [\langle E_i^*(\mathbf{r}, \omega)E_j(\mathbf{r}, \omega)\rangle], \quad (i = x, y; j = x, y). \tag{25}$$

On substituting from Eq. (24) into Eq. (25) we readily find that

$$W_{ij}(\mathbf{r}, \mathbf{r}, \omega) = |K_1|^2 W_{ij}(\rho_1, \rho_1, \omega) + |K_2|^2 W_{ij}(\rho_2, \rho_2, \omega)$$
$$+ K_1^* K_2 W_{ij}(\rho_1, \rho_2, \omega)e^{ik(R_2-R_1)} + K_1 K_2^* W_{ij}(\rho_2, \rho_1, \omega)e^{-ik(R_2-R_1)}, \tag{26}$$

where

$$W_{ij}(\rho_1, \rho_2, \omega) = \langle E_i^*(\rho_1, \omega)E_j(\rho_2, \omega)\rangle, \quad (i = x, y; j = x, y) \tag{27}$$

are the elements of the cross-spectral density matrix whose arguments are the points $Q_1(\rho_1)$ and $Q_2(\rho_2)$ where the pinholes are located.

Suppose that the beam incident on the pinholes is an electromagnetic Gaussian Schell-model beam, which we discussed in Section 9.4.2. The spectral densities $S_j^{(0)}(\rho, \omega)$

---

[1] The analysis of this section is based on a paper by H. Roychowdhury and E. Wolf, *Opt. Commun.* **252** (2005), 268–274.

$(j = x, y)$, of the x and y components of the electric field incident on the pinholes are then given by Eq. (6a) and the correlation coefficients $\mu_{ij}^{(0)}(\rho_2 - \rho_1, \omega)$ are given by Eq. (6b). For simplicity we will assume that the spectral densities of the x and the y components of the electric field vector at the pinholes are the same. Then in Eq. (6a)

$$A_x = A_y \equiv A. \tag{28}$$

In general A will depend on the frequency ω. We will also assume that in Eq. (6b) for the correlation coefficients $\mu_{ij}^{(0)}(\rho_2 - \rho_1, \omega)$,

$$\tag{29a}$$

$$B_{ij} = \begin{cases} 1 & \text{if } i = j, \\ B_0 \text{ (real)} & \text{if } j \neq i, \end{cases} \qquad \begin{matrix} (29a) \\ (29b) \end{matrix}$$

and that

$$\delta_{xx} = \delta_{yy} \equiv \delta. \tag{30}$$

The cross-spectral density matrix of the electric field at the pinholes may then readily be shown to be given by the formula

$$W^{(0)}(\rho_1, \rho_2, \omega) = A^2 \exp\left(-\frac{\rho_1^2 + \rho_2^2}{4\sigma^2}\right) \begin{bmatrix} \exp\left(-\dfrac{(\rho_2 - \rho_1)^2}{2\delta^2}\right) & B_0 \exp\left(-\dfrac{(\rho_2 - \rho_1)^2}{2\delta_{xy}^2}\right) \\ B_0 \exp\left(-\dfrac{(\rho_2 - \rho_1)^2}{2\delta_{xy}^2}\right) & \exp\left(-\dfrac{(\rho_2 - \rho_1)^2}{2\delta_{xy}^2}\right) \end{bmatrix}.$$

$$\tag{31}$$

where we have used the relation $\delta_{yx} = \delta_{xy}$ [Eq. (7d) of Section 9.4.2]. On substituting from Eq. (31) into Eq. (8a) of Section 9.2 one finds that the spectral degree of coherence of the electric field at the two pinholes is given by the expression

$$\eta^{(0)}(\rho_1, \rho_2, \omega) = \exp\left(-\frac{(\rho_2 - \rho_1)^2}{2\delta^2}\right). \tag{32}$$

If the pinholes are placed symmetrically with respect to the z axis, then $\rho_2 = -\rho_1$ and the expression (32) for the spectral degree of coherence becomes

$$\eta^{(0)}(\rho_1, -\rho_1, \omega) = \exp\left(-\frac{2\rho_1^2}{\delta^2}\right). \tag{33}$$

An expression for the degree of polarization at each pinhole is readily obtained on substituting from Eq. (31) into the general expression (14) of Section 9.2. One finds that

$$\mathcal{P}^{(0)}(\rho_\alpha, \omega) \equiv B_0 \quad (\alpha = 1, 2). \tag{34}$$

Since this formula shows that the degree of polarization has the same value, B_0, irrespective of the location of the pinholes, we will suppress the dependence of $\mathcal{P}^{(0)}$ on ρ_α and will write

$$\mathcal{P}^{(0)}(\rho_\alpha, \omega) \equiv \mathcal{P}_0 \, (= B_0). \tag{35}$$

In order to determine the spectral degree of polarization $\mathcal{P}(\mathbf{r}, \omega)$ at a point in the fringe pattern, we need to determine the elements of the matrix $\mathbf{W}$ at a point $P(\mathbf{r})$ in that plane. On substituting from Eq. (31) into Eq. (26), we find that, with the symmetric location of the two pinholes, i.e. with $\rho_2 = -\rho_1$,

$$W_{ij}(\mathbf{r}, \mathbf{r}, \omega) = \begin{cases} A^2 \exp\left(-\dfrac{\rho_1^2}{2\sigma^2}\right)\left[|K_1|^2 + |K_2|^2 + 2\,\mathcal{R}e(K_1^* K_2 e^{ik(R_2 - R_1)})\exp\left(-\dfrac{2\rho_1^2}{\delta^2}\right)\right] \\ \qquad\qquad\qquad \text{when } i = j, \hfill (36a) \\[2ex] A^2 \exp\left(-\dfrac{\rho_1^2}{2\sigma^2}\right)\left[|K_1|^2 + |K_2|^2 + 2\,\mathcal{R}e(K_1^* K_2 e^{ik(R_2 - R_1)})\exp\left(-\dfrac{2\rho_1^2}{\delta_{ij}^2}\right)\right]\mathcal{P}_0 \\ \qquad\qquad\qquad \text{when } i \neq j. \hfill (36b) \end{cases}$$

It is convenient to set

$$S_j(\mathbf{r}, \omega) = |K_j|^2 A^2 \exp\left(-\dfrac{\rho_1^2}{2\sigma^2}\right) \quad (j = 1, 2), \tag{37}$$

which represents the spectral density at frequency ω of the field at the point $\mathbf{r}$ in the plane of observation. The formulas (36) may then be rewritten in the form

$$W_{ij}(\mathbf{r}, \mathbf{r}, \omega)$$

$$= \begin{cases} S_1(\mathbf{r}, \omega) + S_2(\mathbf{r}, \omega) + 2\sqrt{S_1(\mathbf{r}, \omega)}\sqrt{S_2(\mathbf{r}, \omega)}\cos[k(R_2 - R_1)]\exp\left(-\dfrac{2\rho_1^2}{\delta^2}\right) \\ \qquad\qquad\qquad \text{when } i = j, \hfill (38a) \\[2ex] S_1(\mathbf{r}, \omega) + S_2(\mathbf{r}, \omega) + 2\sqrt{S_1(\mathbf{r}, \omega)}\sqrt{S_2(\mathbf{r}, \omega)}\cos[k(R_2 - R_1)]\exp\left(-\dfrac{2\rho_1^2}{\delta_{ij}^2}\right)\mathcal{P}_0 \\ \qquad\qquad\qquad \text{when } i \neq j. \hfill (38b) \end{cases}$$

On substituting from Eqs. (38) into Eq. (14) of Section 9.2 we obtain the following expression for the degree of polarization in the plane $\mathcal{B}$ of the interference pattern:

$$\mathcal{P}(\mathbf{r}, \omega)$$

$$= \frac{S_1(\mathbf{r}, \omega) + S_2(\mathbf{r}, \omega) + 2\sqrt{S_1(\mathbf{r}, \omega)}\sqrt{S_2(\mathbf{r}, \omega)}\cos[k(R_2 - R_1)]\exp\left(-\dfrac{2\rho_1^2}{\delta_{xy}^2}\right)}{S_1(\mathbf{r}, \omega) + S_2(\mathbf{r}, \omega) + 2\sqrt{S_1(\mathbf{r}, \omega)}\sqrt{S_2(\mathbf{r}, \omega)}\cos[k(R_2 - R_1)]\exp\left(-\dfrac{2\rho_1^2}{\delta^2}\right)}\mathcal{P}_0.$$

$$\tag{39}$$

According to Eq. (33) the exponential factor in the last term in the denominator is just the spectral degree of coherence $\eta^{(0)}(\rho_1, -\rho_1, \omega)$ of the light at the two pinholes. Hence Eq. (39) may be rewritten in the form

$$\mathcal{P}(\mathbf{r}, \omega)$$

$$= \frac{S_1(\mathbf{r}, \omega) + S_2(\mathbf{r}, \omega) + 2\sqrt{S_1(\mathbf{r}, \omega)}\sqrt{S_2(\mathbf{r}, \omega)}\cos[k(R_2 - R_1)]\exp\left(-\frac{2\rho_1^2}{\delta_{xy}^2}\right)}{S_1(\mathbf{r}, \omega) + S_2(\mathbf{r}, \omega) + 2\sqrt{S_1(\mathbf{r}, \omega)}\sqrt{S_2(\mathbf{r}, \omega)}\cos[k(R_2 - R_1)]\eta^{(0)}(\rho_1, -\rho_1, \omega)}\mathcal{P}_0.$$

(40)

Frequently the spectral densities $S_1(\mathbf{r}, \omega)$ and $S_2(\mathbf{r}, \omega)$ of the light reaching the point $P(\mathbf{r})$ in the plane $\mathcal{B}$ will be approximately the same, i.e. $S_2(\mathbf{r}, \omega) \approx S_1(\mathbf{r}, \omega)$. Equation (40) then reduces to

$$\mathcal{P}(\mathbf{r}, \omega) = \frac{1 + \cos[k(R_2 - R_1)]\exp\left(-\frac{2\rho_1^2}{\delta_{xy}^2}\right)}{1 + \cos[k(R_2 - R_1)]\eta^{(0)}(\rho_1, -\rho_1, \omega)}\mathcal{P}_0.$$

(41)

This formula shows that *the degree of polarization in the detection plane $\mathcal{B}$ depends not only on the degree of polarization $\mathcal{P}_0$ of the light at the pinholes but also on the spectral degree of coherence $\eta^{(0)}$ of the electric field at the pinholes and on the parameter δ_{xy} which characterizes the correlation between the x component of the electric field at one of the pinholes and the y component of the electric field at the other pinhole*. This result brings into evidence the subtle relations which exist between polarization and coherence of stochastic electromagnetic beams.

Figure 9.7 shows the behavior of the spectral degree of polarization $\mathcal{P}$ at the axial point[1] (i.e. on the beam axis) in the plane of observation $\mathcal{B}$, when the spectral degree of coherence $\eta^{(0)}(\rho_1, -\rho_1, \omega)$ at the pinholes is varied while the degree of polarization $\mathcal{P}_0$ at each pinhole is kept constant.

The effect of the degree of coherence of light at the pinholes on the degree of polarization in Young's interference experiment was tested and confirmed experimentally.[2]

9.5 Generalized Stokes parameters[3]

In Section 8.1 we introduced the Stokes parameters of a random electromagnetic beam. They were defined in terms of the mutual intensity matrix $\mathbf{J}$ by Eq. (14a) of that section, i.e.

[1] The behavior of the spectral degree of polarization at off-axis points is discussed in Y. Li, H. Lee and E. Wolf, *Opt. Commun.* **265** (2006), 63–72.

[2] F. Gori, M. Santarsiero, R. Borghi and E. Wolf, *Opt. Lett.* **31** (2006), 688–690.

[3] The results presented in this section are based on a paper by O. Korotkova and E. Wolf, *Opt. Lett.* **30** (2005), 198–200.

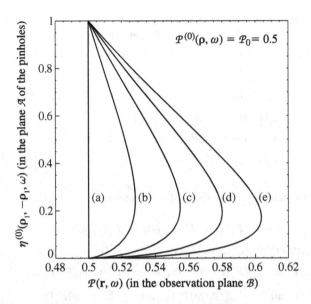

Fig. 9.7 Behavior of the spectral degree of polarization $\mathcal{P}(\mathbf{r}, \omega)$, at the axial point in the plane of observation $\mathcal{B}$, when the spectral degree of coherence $\eta^{(0)}(\rho_1, -\rho_1, \omega)$ of the light at each pinhole is changed, while the degree of polarization $\mathcal{P}_0$ at each pinhole is kept fixed. The curves are associated with different values of the parameter δ_{xy} which characterizes the correlation between the x component of the electric field at one pinhole and the y component of the electric field at the other pinhole. The values of the parameters were taken as $\mathcal{P}_0 = 0.5$, $z = 10\,\text{cm}$, $\rho_1 = 2\,\text{mm}$,

(a) $\delta_{xy} = \delta$,

(b) $\delta_{xy} = \dfrac{\delta}{4}\left(3 + \dfrac{1}{\sqrt{\mathcal{P}_0}}\right) = 1.1\delta$,

(c) $\delta_{xy} = \dfrac{\delta}{2}\left(3 + \dfrac{1}{\sqrt{\mathcal{P}_0}}\right) = 1.2\delta$,

(d) $\delta_{xy} = \dfrac{\delta}{4}\left(1 + \dfrac{3}{\sqrt{\mathcal{P}_0}}\right) = 1.3\delta$,

(e) $\delta_{xy} = \dfrac{\delta}{\sqrt{\mathcal{P}_0}} = 1.4\delta$.

[After H. Roychowdhury and E. Wolf, *Opt. Commun.* **252** (2005), 268–274.]

in terms of equal-time correlations between mutually orthogonal components of the electric field at one point. It is also possible and useful to define analogous quantities in the space–frequency domain by somewhat similar formulas, in terms of the cross-spectral density matrix, viz.,

$$
\left.
\begin{aligned}
s_0(\mathbf{r}, \omega) &= W_{xx}(\mathbf{r}, \mathbf{r}, \omega) + W_{yy}(\mathbf{r}, \mathbf{r}, \omega), \\
s_1(\mathbf{r}, \omega) &= W_{xx}(\mathbf{r}, \mathbf{r}, \omega) - W_{yy}(\mathbf{r}, \mathbf{r}, \omega), \\
s_2(\mathbf{r}, \omega) &= W_{xy}(\mathbf{r}, \mathbf{r}, \omega) + W_{yx}(\mathbf{r}, \mathbf{r}, \omega), \\
s_3(\mathbf{r}, \omega) &= i\left[W_{yx}(\mathbf{r}, \mathbf{r}, \omega) - W_{xy}(\mathbf{r}, \mathbf{r}, \omega)\right],
\end{aligned}
\right\}
\tag{1a}
$$

or, more explicitly, in terms of the spectral components of the electric field,

$$
\left.
\begin{aligned}
s_0(\mathbf{r}, \omega) &= \langle E_x^*(\mathbf{r}, \omega)E_x(\mathbf{r}, \omega)\rangle + \langle E_y^*(\mathbf{r}, \omega)E_y(\mathbf{r}, \omega)\rangle, \\
s_1(\mathbf{r}, \omega) &= \langle E_x^*(\mathbf{r}, \omega)E_x(\mathbf{r}, \omega)\rangle - \langle E_y^*(\mathbf{r}, \omega)E_y(\mathbf{r}, \omega)\rangle, \\
s_2(\mathbf{r}, \omega) &= \langle E_x^*(\mathbf{r}, \omega)E_y(\mathbf{r}, \omega)\rangle + \langle E_y^*(\mathbf{r}, \omega)E_x(\mathbf{r}, \omega)\rangle, \\
s_3(\mathbf{r}, \omega) &= i\left[\langle E_y^*(\mathbf{r}, \omega)E_x(\mathbf{r}, \omega)\rangle - \langle E_x^*(\mathbf{r}, \omega)E_y(\mathbf{r}, \omega)\rangle\right].
\end{aligned}
\right\}
\tag{1b}
$$

These four parameters may be called the *spectral Stokes parameters*. They can be determined experimentally in a similar way to how one determines the usual Stokes parameters, provided that the light is filtered to become quasi-monochromatic around the frequency ω.

In Section 9.4.1 we showed how the cross-spectral density matrix of a stochastic electromagnetic beam changes on propagation. In this section we will briefly consider how the spectral Stokes parameters change as the beam propagates. For this purpose it is, however, necessary to generalize the usual Stokes parameters, which depend on one point, to quantities which depend on two points. Such generalized Stokes parameters may be introduced by formulas of the form given by Eq. (1), but with the two equal spatial arguments $(\mathbf{r}, \mathbf{r})$ on the right-hand side replaced by two unequal spatial arguments $(\mathbf{r}_1, \mathbf{r}_2)$. Thus, in place of the formulas (1a) we now have

$$
\left.
\begin{aligned}
S_0(\mathbf{r}_1, \mathbf{r}_2, \omega) &= W_{xx}(\mathbf{r}_1, \mathbf{r}_2, \omega) + W_{yy}(\mathbf{r}_1, \mathbf{r}_2, \omega), \\
S_1(\mathbf{r}_1, \mathbf{r}_2, \omega) &= W_{xx}(\mathbf{r}_1, \mathbf{r}_2, \omega) - W_{yy}(\mathbf{r}_1, \mathbf{r}_2, \omega), \\
S_2(\mathbf{r}_1, \mathbf{r}_2, \omega) &= W_{xy}(\mathbf{r}_1, \mathbf{r}_2, \omega) + W_{yx}(\mathbf{r}_1, \mathbf{r}_2, \omega), \\
S_3(\mathbf{r}_1, \mathbf{r}_2, \omega) &= i\left[W_{yx}(\mathbf{r}_1, \mathbf{r}_2, \omega) - W_{xy}(\mathbf{r}_1, \mathbf{r}_2, \omega)\right].
\end{aligned}
\right\}
\tag{2a}
$$

More explicitly,

$$
\left.
\begin{aligned}
S_0(\mathbf{r}_1, \mathbf{r}_2, \omega) &= \langle E_x^*(\mathbf{r}_1, \omega)E_x(\mathbf{r}_2, \omega)\rangle + \langle E_y^*(\mathbf{r}_1, \omega)E_y(\mathbf{r}_2, \omega)\rangle, \\
S_1(\mathbf{r}_1, \mathbf{r}_2, \omega) &= \langle E_x^*(\mathbf{r}_1, \omega)E_x(\mathbf{r}_2, \omega)\rangle - \langle E_y^*(\mathbf{r}_1, \omega)E_y(\mathbf{r}_2, \omega)\rangle, \\
S_2(\mathbf{r}_1, \mathbf{r}_2, \omega) &= \langle E_x^*(\mathbf{r}_1, \omega)E_y(\mathbf{r}_2, \omega)\rangle + \langle E_y^*(\mathbf{r}_1, \omega)E_x(\mathbf{r}_2, \omega)\rangle, \\
S_3(\mathbf{r}_1, \mathbf{r}_2, \omega) &= i\left[\langle E_y^*(\mathbf{r}_1, \omega)E_x(\mathbf{r}_2, \omega)\rangle - \langle E_x^*(\mathbf{r}_1, \omega)E_y(\mathbf{r}_2, \omega)\rangle\right].
\end{aligned}
\right\}
\tag{2b}
$$

If the beam propagates in free space, the cross-spectral density matrix $W(\mathbf{r}_1, \mathbf{r}_2, \omega)$ propagates from an initial plane $z = 0$ according to Eq. (3) of Section 9.4.1. It follows at once from Eqs. (2a) for the generalized spectral Stokes parameters that these parameters propagate according to the same law, viz.,

$$S_\alpha(\mathbf{r}_1, \mathbf{r}_2, \omega)$$
$$= \iint_{z=0} S_\alpha^{(0)}(\rho_1', \rho_2', \omega) K(\rho_1 - \rho_1', \rho_2 - \rho_2', z; \omega) d^2\rho_1' \, d^2\rho_2', \quad (\alpha = 0, 1, 2, 3),$$

(3)

where $\mathbf{r}_1 = (\rho_1, z)$; $\mathbf{r}_2 = (\rho_2, z)$ and the propagator $K(\rho_1 - \rho_1', \rho_2 - \rho_2', z; \omega)$ is given by Eqs. (4) and (2) of Section 9.4.1.

If we set $\mathbf{r}_1 = \mathbf{r}_2 \equiv \mathbf{r}$ in Eq. (3) the formula becomes, with $\mathbf{r} \equiv (\rho, z)$,

$$s_\alpha(\mathbf{r}, \omega) = S_\alpha(\mathbf{r}, \mathbf{r}, \omega) = \iint_{z=0} S_\alpha^{(0)}(\rho_1', \rho_2', \omega) K(\rho - \rho_1', \rho - \rho_2', z; \omega) d^2\rho_1' \, d^2\rho_2'. \quad (4)$$

This formula expresses the ordinary spectral Stokes parameters $s_\alpha(\mathbf{r}, \omega)$ at any point in the half-space $z > 0$ in terms of the generalized spectral Stokes parameters S_α at all pairs of points in the plane $z = 0$. Evidently knowledge of the ordinary spectral Stokes parameters $s_\alpha(\mathbf{r}, \omega)$ at all points in the source plane is not adequate to determine the values of those parameters throughout the half-space $z > 0$. In fact two stochastic electromagnetic beams which have the same Stokes parameters $s_\alpha(\rho, \omega)$ in the source plane $z = 0$ may produce beams of different degrees of polarization throughout the half-space $z > 0$.[1]

It is clear that the generalized spectral Stokes parameters, just like the cross-spectral density matrix, can be used to elucidate the spectral properties, the polarization properties and the coherence properties of a stochastic electromagnetic beam.

In Figure 9.8 examples are given of the changes of the spectral Stokes parameters of an electromagnetic Gaussian Schell-model beam on propagation in free space, calculated from Eqs. (4) and (2a).

Similar calculations can be carried out for propagation of the spectral Stokes parameters not only in free space but in any linear medium, deterministic or random. One must then use in Eq. (4) the kernel appropriate to propagation in that medium.

PROBLEMS

9.1 Two unpolarized, mutually uncorrelated stochastic, statistically stationary beams propagate close to the z axis into the half-space $z > 0$. $\{\mathbf{E}^{(A)}(\mathbf{r}, \omega)\}$ and $\{\mathbf{E}^{(B)}(\mathbf{r}, \omega)\}$ are the statistical ensembles representing each beam. Further,

$$W_{ij}^{(A)} = \langle E_i^{(A)*} E_j^{(A)} \rangle, \qquad W_{ij}^{(B)} = \langle E_i^{(B)*} E_j^{(B)} \rangle$$

---

[1] An example of this kind is given in M. Salem, O. Korotkova and E. Wolf, *Opt. Lett.* **31** (2006), 3025–3027, where the difference in the degrees of polarization of the two beams is due to the difference in the degrees of coherence of the field in the source plane $z = 0$, information which is not contained in the ordinary Stokes parameters s_j.

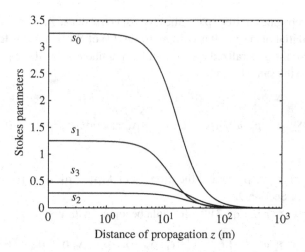

Fig. 9.8 Changes of the Stokes parameters s_0, s_1, s_2 and s_3 of an electromagnetic Gaussian Schell-model beam on propagation in free space. The parameters characterizing the source have been taken as $\omega = 3 \times 10^{15}$ Hz ($\lambda = 6328$ nm), $A_x = 1.5$, $A_y = 1$, $\delta = \arg B_{xy} = \pi/6$, $|B_{xy}| = 0.35$, $\sigma = 1$ cm, $\delta_{yy} = 0.2$ mm, $\delta_{xy} = 0.25$ mm and $\delta_{xx} = 0.15$ mm. [Reproduced from O. Korotkova and E. Wolf, *Opt. Lett.* **30** (2005), 198–200.]

and

$$W_{ij}^{(A,B)} = \langle E_i^{(A)*} E_j^{(B)} \rangle, \qquad (i = x, y; \ j = x, y),$$

x and y being mutually orthogonal directions perpendicular to the z direction. The angular brackets denote the ensemble average.

(a) Derive an expression for the cross-spectral density matrix of the total field.

(b) Show that the total field may be partially polarized. Under what condition will the total field be unpolarized?

9.2 A thermal light beam, with axis along the z direction, is incident on two detectors located at points $P_1(\rho_1)$ and $P_2(\rho_2)$ in a plane $z =$ constant. Assuming that the 2×2 cross-spectral density of the beam is symmetric in the variables ρ_1 and ρ_2, show that the correlation between the intensity fluctuations $\Delta I(\rho_j, \omega) = I(\rho_j, \omega) - \langle I(\rho_j, \omega) \rangle$ ($j = 1, 2$) is given by the expression

$$\langle \Delta I(\rho_1, \omega) \Delta I(\rho_2, \omega) \rangle = \mathrm{Tr}[W^\dagger(\rho_1, \rho_2, \omega) W(\rho_1, \rho_2, \omega)],$$

where $W(\rho_1, \rho_2, \omega)$ is the cross-spectral density matrix of the beam, and the dagger denotes the Hermitian adjoint.

9.3 Derive an expression for the degree of coherence of a planar, secondary, electromagnetic Gaussian Schell-model source.

9.4 Show that the degree of coherence at any pair of points in a cross-section of an electro-magnetic Gaussian Schell-model beam tends to unity with increasing distance of propagation in free space.

9.5 Consider a beam generated by a Gaussian Schell-model source located in the plane $z = 0$ and linearly polarized along the x direction, with spectral densities

$$S_x^{(0)}(\omega) = A^2 e^{-(\omega-\bar{\omega})^2/(2\sigma^2)}, \qquad S_y^{(0)}(\omega) = B^2 e^{-(\omega-\bar{\omega})^2/(2\sigma^2)}$$

and correlation coefficients of the electric field

$$\mu_{xx}^{(0)}(\rho_1, \rho_2, \omega) = e^{-(\rho_1-\rho_2)^2/(2\delta_x^2)},$$
$$\mu_{yy}(\rho_1, \rho_2, \omega) = e^{-(\rho_1-\rho_2)^2/(2\delta_y^2)},$$
$$\mu_{xy}(\rho_1, \rho_2, \omega) = 0.$$

In these expressions A, B, σ, δ_x and δ_y are independent of position and of frequency and the source is assumed to be large relative to σ. Let

$$s(\mathbf{r}, \omega) = \frac{S(\mathbf{r}, \omega)}{\int_0^\infty S(\mathbf{r}, \omega)\,d\omega},$$

be the normalize spectral density at the point $\mathbf{r}$. Derive an expression for $s(\mathbf{r}, \omega)$ at points in the far zone.

Plot the normalized source spectrum and the normalized far-zone spectrum in directions making angles $\theta = 0°$ and $\theta = 0.3°$ with the normal to the source plane, when $A = 1$, $\sigma = 0.2\bar{\omega}$ and $\delta = 0.5\,\text{mm}$.

Determine and plot also the degree of coherence $\eta^{(0)}(\rho_1, \rho_2, \omega)$ of the electric field in the source plane at pairs of points located symmetrically with respect to the beam axis, as a function of the separation $|\Delta\rho| = 2|\rho_1| = 2|\rho_2|$.

9.6 Consider a stochastic electromagnetic beam generated by a planar, secondary source, located in the plane $z = 0$ and propagating about the z-axis into the half-space $z > 0$. The spectral densities $S_x^{(0)}(\omega)$ and $S_y^{(0)}(\omega)$ of two mutually orthogonal components of the electric field in the source plane are assumed to be independent of position.

Show that the spectral density of the field at a point $\mathbf{r}$ in the half-space $z > 0$ may be expressed in the form

$$S(\mathbf{r}, \omega) = S_x^{(0)}(\omega) M_x(\mathbf{r}, \omega) + S_y^{(0)}(\omega) M_y(\mathbf{r}, \omega).$$

Derive expressions for $M_x(\mathbf{r}, \omega)$ and $M_y(\mathbf{r}, \omega)$. How do these expressions simplify when the point $\mathbf{r}$ is in the far zone?

Show that, if the components E_x and E_y of the electric field in the source plane are uncorrelated, the above formula can be expressed in the form

$$S(\mathbf{r}, \omega) = S_x^{(0)}(\omega)[M_x(\mathbf{r}, \omega) + \alpha(\mathbf{r}, \omega)M_y(\mathbf{r}, \omega)].$$

where

$$\alpha(\omega) = \frac{1 - \mathcal{P}^{(0)}(\omega)}{1 + \mathcal{P}^{(0)}(\omega)},$$

$\mathcal{P}^{(0)}(\omega)$ being the degree of polarization of the field in the source plane.

9.7 σ_0 denotes the unit matrix and σ_1, σ_2 and σ_3 are the so-called Pauli spin matrices. The four matrices are defined as

$$\sigma_0 = \begin{bmatrix} 1 & 0 \\ 0 & 1 \end{bmatrix}, \qquad \sigma_1 = \begin{bmatrix} 1 & 0 \\ 0 & -1 \end{bmatrix}, \qquad \sigma_2 = \begin{bmatrix} 0 & 1 \\ 1 & 0 \end{bmatrix}, \qquad \sigma_3 = \begin{bmatrix} 0 & i \\ -i & 0 \end{bmatrix}.$$

(a) Show that

$$\mathrm{Tr}(\sigma_j \cdot \sigma_k) = 2\delta_{jk} \qquad (j, k = 0, 1, 2, 3),$$

where δ_{jk} is the Kronecker symbol ($\delta_{jk} = 1$ when $k = j$; $\delta_{jk} = 0$ when $k \neq j$).

(b) Show with the help of the above relations that the spectral Stokes parameters and the cross-spectral density matrix are related by the formulas

$$s_j(\mathbf{r}, \omega) = \mathrm{Tr}\{W(\mathbf{r}, \mathbf{r}, \omega) \cdot \vec{\sigma}_j\}$$

and

$$W(\mathbf{r}, \mathbf{r}, \omega) = \frac{1}{2}\sum_{i=1}^{3}\{s_j(\mathbf{r}, \omega)\vec{\sigma}_j\}.$$

9.8 $s_j(\mathbf{r}, \omega)$ ($j = 0, 1, 2, 3$) are the spectral Stokes parameters of a quasi-monochromatic, stationary light beam. The matrix

$$S(\mathbf{r}, \omega) \equiv \begin{bmatrix} s_0(\mathbf{r}, \omega) \\ s_1(\mathbf{r}, \omega) \\ s_2(\mathbf{r}, \omega) \\ s_3(\mathbf{r}, \omega) \end{bmatrix}$$

is said to represent the (spectral) Stokes vector of the beam.

Suppose that the beam passes through a linear, non-image-forming device. Let $\mathbf{S}'(\mathbf{r}, \omega)$ denote the Stokes vector of the emerging beam. Then

$$\mathbf{S}'(\mathbf{r}, \omega) = \mathbf{M}(\mathbf{r}, \omega)\mathbf{S}(\mathbf{r}, \omega),$$

where M is a 4×4 matrix known as the (spectral) Mueller matrix.

Express the elements of the Mueller matrix in terms of the elements of the instrumental matrix, defined in Problem 8.6.

9.9 Show that the two stochastic electromagnetic beams which propagate from the source plane $z = 0$ into the half-space $z > 0$ may have different degrees of polarization throughout the half-space, even though they have the same sets of Stokes parameters in the source plane. (**Hint:** use the generalized Stokes parameters to characterize the source.)

Appendix I

Cells of phase space and
the degeneracy parameter

(a) Cells of phase space of a quasi-monochromatic light wave (Section 1.4)

We mentioned towards the end of Section 1.4 that the concept of coherence volume has a counterpart in the quantum theory of radiation, known as a *cell of phase space*. In this appendix we will define the cell of phase space and also introduce a related concept of degeneracy parameters of radiation.

Let us first consider a monochromatic plane wave of wavelength λ. According to one of the de Broglie relations (B&W, Appendix II), we may associate with it photons of momentum $\mathbf{p}$, whose magnitude is

$$p = \frac{h}{\lambda}, \tag{1}$$

where h is Planck's constant. Suppose that $\mathbf{r} \equiv (x, y, z)$ specifies the location of a photon.[1] According to elementary quantum mechanics, the location and the momentum of a photon cannot be measured simultaneously with an accuracy greater than allowed by the Heisenberg uncertainty relation, viz.,[2]

$$\Delta x \, \Delta p_x \geq h, \qquad \Delta y \, \Delta p_y \geq h, \qquad \Delta z \, \Delta p_z \geq h. \tag{2}$$

Let us introduce a six-dimensional space, called the *phase space*, in which points are specified by values of the six variables x, y, z, p_x, p_y, p_z. In view of the inequalities (2) it is natural to divide this space into regions of volume elements, called *cells*, of size

$$\Delta x \, \Delta y \, \Delta z \, \Delta p_x \Delta p_y \Delta p_z = h^3. \tag{3}$$

Clearly photons of the same spin state (same polarization) belonging to a region of the phase space which is not greater than indicated by Eq. (3) are *intrinsically indistinguishable*.

---

[1] The position of a photon cannot be specified more accurately than to distances of the order of the wavelength but here we will ignore subtleties concerning photon localization.

[2] The uncertainty relations (2) are often written with a factor $\hbar/2$ ($\hbar = h/(2\pi)$), rather than h on the right-hand sides. The value of the factor depends on the exact definitions of the uncertainties Δx, Δp_x, etc. For our purposes definitions which imply the inequalities (2) are convenient.

We will now show that the concept of coherence volume which we introduced in Section 1.4 from considerations based on classical wave theory is the volume $\Delta x \, \Delta y \, \Delta z$ of ordinary space given by the formula (3), subject to the constraints imposed on the product $\Delta p_x \, \Delta p_y \, \Delta p_z$ by the geometry and by the bandwidth of the light. Stated somewhat differently, we will show that the *coherence volume is that region of space throughout which the photons are intrinsically indistinguishable.* To justify this statement let us first estimate the uncertainty in the components of the momentum of a photon in the far zone of a field generated by a planar, quasi-monochromatic, thermal source σ, of *linear dimensions* $2a$ located in the plane $z = 0$ and radiating into the half-space $z > 0$. Let 2ϕ denote the angle which the source subtends at a point Q in the far zone, assumed, for simplicity, to be small and situated on the normal to σ, at a distance R from it (Fig. 1). There will be uncertainties Δp_x and Δp_y in the x and the y components of the momentum of the photon arriving at Q, because of the lack of knowledge regarding the source point from which the photon was emitted. Evidently these uncertainties are represented by the projections of the photon momentum $\mathbf{p}$ onto the x and the y axes, i.e.

$$\Delta p_x = \Delta p_y \approx 2p\phi. \tag{4}$$

Using Eq. (1) and assuming ϕ to be sufficiently small, one has

$$\begin{aligned}
\Delta p_x = \Delta p_y &\approx 2p\phi \\
&= \frac{2h}{\bar{\lambda}} \phi \approx \frac{2h}{\bar{\lambda}} \frac{a}{R},
\end{aligned} \tag{5}$$

where $\bar{\lambda}$ is the mean wavelength of the emitted light. Since ϕ was assumed to be small, the uncertainty of the z component of the momentum arises mainly from the uncertainty in the wavelength. If $\Delta\lambda$ is the effective wavelength range of the light, then it follows from Eq. (1) that

$$\Delta p_z = \frac{h}{\bar{\lambda}^2} \Delta\lambda. \tag{6}$$

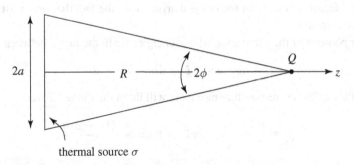

thermal source σ

Fig. 1 Notation relating to the derivation of Eq. (5) of Appendix I.

Hence we have from Eqs. (5) and (6)

$$\Delta p_x \, \Delta p_y \, \Delta p_z = h^3 \frac{\Delta \lambda}{\overline{\lambda}^4} \frac{S}{R^2}, \tag{7}$$

where $S = (2a)^2$ is the order of magnitude of the size of the source. On substituting this expression into the expression (3) for a cell of phase space, we see that the volume of the space around the point Q throughout which the photons emitted by the source are intrinsically indistinguishable is

$$\Delta x \, \Delta y \, \Delta z = \frac{R^2}{S} \left(\frac{\overline{\lambda}}{\Delta \lambda} \right) \overline{\lambda}^3. \tag{8}$$

Comparison of Eq. (8) with Eq. (2a) of Section 1.4 shows that the right-hand side is equal to the expression for the coherence volume derived by considerations based entirely on classical theory. Thus we have justified our earlier assertion about the quantum-mechanical significance of the coherence volume.

(b) Cells of phase space of radiation in a cavity (Sections 7.4 and 7.5)

The concept of cells of phase space was originally introduced not for a light beam but rather for thermal radiation in a thermally insulated cavity, i.e. for blackbody radiation. It plays an important role in the theory of photoelectric detection of light fluctuations, for example, which is discussed in Section 7.5. In this appendix we will derive an expression for the number of cells of phase space for this situation.

Using Eq. (3) let us calculate the number Z, say, of cells of phase space associated with photons in some finite momentum range contained in a volume V of the cavity. Evidently

$$Z = \frac{2}{h^3} \int dp_x \, dp_y \, dp_z \, dx \, dy \, dz, \tag{9}$$

where the integration extends over the accessible domain of the phase space containing the photons. The factor 2 in front of the integral arises from the fact that one must distinguish between two spin states.

Consider photons of the same spin, whose energies lie in the range between

$$E = h\nu \quad \text{and} \quad E + dE = h(\nu + d\nu).$$

The magnitude of the corresponding momenta will lie in the range between

$$p = \frac{E}{c} = \frac{h\nu}{c} \quad \text{and} \quad p + dp = \frac{h(\nu + d\nu)}{c},$$

where c is the speed of light in vacuum.

Since we are considering radiation under equilibrium conditions, there is no preferential direction of propagation of the photons. This makes it possible to determine the contribution to the integral in Eq. (9) from the momenta in a simple manner as follows. Consider a spherical shell in momentum space indicated on the figure below:

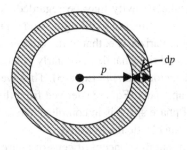

The contribution to Eq. (9) from the shell is evidently

$$\int dp_x \, dp_y \, dp_z = 4\pi p^2 \, dp$$

$$= 4\pi \left(\frac{h\nu}{c}\right)^2 \frac{h}{c} \, d\nu$$

$$= 4\pi \frac{h^3}{c^3} \nu^2 \, d\nu. \tag{10}$$

The contribution from ordinary space (called also configuration space) is

$$\int dx \, dy \, dz = V. \tag{11}$$

On substituting from Eqs. (10) and (11) into Eq. (9) we obtain for the number Z of cells in phase space the expression

$$Z = \frac{2}{h^3} 4\pi \frac{h^3}{c^3} \nu^2 \, d\nu \, V,$$

i.e.

$$Z = \frac{8\pi V}{c^3} \nu^2 \, d\nu. \tag{12}$$

The size of the region in the cavity throughout which the photons are indistinguishable is, according to Eq. (12) with $Z = 1$, given by the expression

$$V = \frac{1}{8\pi} \left(\frac{\lambda}{\Delta\lambda}\right) \lambda^3, \tag{13}$$

where we have used the relation $\nu = c/\lambda$. Equation (13) is of the form of Eq. (2b) of Section 1.4 for the coherence volume of thermal light, with the solid angle $\Delta\Omega'$ replaced by 8π. Now $\Delta\Omega'$ represents the solid angle formed by all the directions along which radiation in the cavity can reach the region under consideration. As was mentioned in Section 7.4, Einstein pointed out in a basic paper on energy fluctuations in blackbody radiation that radiation in a thermally insulated cavity may be regarded as a mixture of plane waves which propagate in all possible directions, filling a solid angle 4π. If we also take into account that the waves are unpolarized, i.e. that each consists of two mutually independent polarization states (e.g. left- and right-handed circularly polarized waves), one can at once understand the origin of the factor $1/(8\pi)$ in Eq. (13). Thus we see that, just as in the case discussed in part (a) of this appendix, Eq. (13) derived from the concept of indistinguishability of photons in a cell of phase space is in complete agreement with the expression for the coherence volume based on classical wave theory.

Finally we might mention that the concept of cells of phase space is often encountered in the physics literature in various disguises. For example, one sometimes speaks of an *elementary bundle of rays* (a concept due to von Laue), which, just like the term "volume of an oscillation mode," means radiation occupying a single cell of phase space. Sometimes one finds a reference to the *degrees of freedom of a light beam* or its *Jean's number*. These are merely alternative terms for the number of cells of phase space.[1]

(c) The degeneracy parameter

For the analysis of some situations, encountered, for example, in the theory of photoelectric detection of light fluctuations (Section 7.5), it is important to have an estimate for the average number of photons contained in a cell of phase space. This quantity is known as the *degeneracy parameter* of the radiation. It differs drastically for thermal and for laser light, as we will now show.

For blackbody radiation at equilibrium temperature T, the value of the degeneracy parameter, δ, at frequency ν, is given by the formula[2]

$$\delta = \frac{1}{e^{h\nu/(k_B T)} - 1},\tag{14}$$

k_B being the Boltzman constant. Figure 2 indicates the values of the degeneracy parameter of blackbody radiation for various temperatures and wavelengths $\lambda = c/\nu$, calculated from Eq. (14). For light of frequency $\nu = 5 \times 10^{14}$ Hz ($\lambda = 6 \times 10^{-5}$ cm) from an incandescent source at temperature $T = 3,000$ K,

$$\delta \sim 3 \times 10^{-4},\tag{15}$$

---

[1] In this connection see also D. Gabor in *Progress in Optics*, Vol. I, E. Wolf ed. (North-Holland, Amsterdam, 1961), Section V, p. 146 *et seq.*
[2] See L. Mandel, *J. Opt. Soc. Amer.* **51** (1961), 797–798.

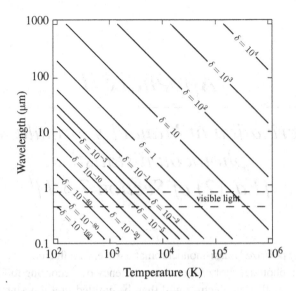

Fig. 2 The degeneracy parameters of blackbody radiation for various temperatures and wavelengths. [Reproduced from W. Martienssen and E. Spiller, *Amer. J. Phys.* **32** (1964), 919–926.

indicating that such light is highly non-degenerate ($\delta \ll 1$). In order that $\delta \sim 1$ at this frequency, it is necessary that the temperature be of the order of 3×10^4K.

The situation is quite different for laser light. To see this, let us consider a laser with 1 mW output power, generating a beam of cross-section $1\,\text{mm}^2$ and having mean wavelength $\bar{\lambda} = 6 \times 10^{-5}$cm ($\bar{\nu} = 5 \times 10^{14}$Hz). The number of photons per unit volume expressed in terms of the energy of a single photon in such a light beam is

$$\rho = \frac{\text{photons}}{\text{volume}} = \frac{\text{(beam power)}}{(h\nu)(c)(\text{cross-sectional area of beam})}$$

$$= \frac{(1 \times 10^{-3}\,\text{J/s})}{(3.3 \times 10^{-19}\,\text{J})(3 \times 10^{8}\,\text{m/s})(1 \times 10^{-6}\,\text{m}^2)}$$

$$= 10^{13}\,\text{photons/m}^3$$

$$= 10^{7}\,\text{photons/cm}^3. \tag{16}$$

We have seen earlier [Eq. (7) of Section 1.4] that, over a short enough time interval, such a laser is sufficiently stable that the coherence volume of the light which it generates is $\Delta V \sim 300\,\text{cm}^3$. Hence, in this case, the degeneracy parameter has the value

$$\delta = \rho \cdot \Delta V \sim (10^7\,\text{photons/cm}^3) \times 300\,\text{cm}^3 = 3 \times 10^9. \tag{17}$$

Such light is, therefore, highly degenerate ($\delta \gg 1$). Comparison of Eqs. (17) and (15) shows the remarkable difference of 13 orders of magnitude between the degeneracy of blackbody radiation and that of the laser beam.

Appendix II

Derivation of Mandel's formula for photocount statistics
[Eq. (2) of Section 7.5.1][1]

Consider a linearly polarized quasi-monochromatic wave, with fluctuating complex amplitude $V(t)$, incident on a photodetector of quantum efficiency α. According to Eq. (1) of Section 7.5.1, the probability that an electron will then be emitted in a time-interval $(t, t + T)$ is given by $P(t)\Delta t = \alpha I(t)\Delta t$.

Let us divide the interval $(t, t + T)$ into $T/\Delta t$ short intervals, each of duration Δt, which we will label as

$$t_i = t + i\,\Delta t \quad (i = 0, 1, 2, \ldots, T/\Delta t). \tag{1}$$

The probability of obtaining n counts in the interval $(t, t + T)$ is the product of the probabilities of obtaining a count at time t_{r_1}, a count at time $t_{r_2}, \ldots$, a count at time t_{r_n}, multiplied by the probability of obtaining no count in the remaining $(T/\Delta t) - n$ intervals, summed over all possible sequences of the counts. Thus

$$p(n, t, T) = \lim_{\Delta t \to 0} \sum_{r_1=0}^{T/\Delta t} \sum_{r_2=0}^{T/\Delta t} \cdots \sum_{r_n=0}^{T/\Delta t} \frac{1}{n!} \alpha^n I(t_{r_1}) I(t_{r_2}) \cdots I(t_{r_n})(\Delta t)^n$$

$$\times \prod_{i=0}^{T/\Delta t} [1 - \alpha I(t_i)\Delta T] \bigg/ \prod_{j=1}^{n} [1 - \alpha I(t_{r_j})\Delta t]. \tag{2}$$

We note that as $\Delta t \to 0$ the product

$$\prod_{j=1}^{n} [1 - \alpha I(t_{r_j})\Delta t] \to 1 - nO(\Delta t), \tag{3}$$

---

[1] The analysis of this appendix follows essentially a derivation given by L. Mandel in *Progress in Optics*, Vol. II, E. Wolf ed. (North-Holland, Amsterdam, 1963), pp. 242–248.

which evidently tends to unity if n is finite. The multiple summation over the indices $r_1, r_2, \ldots, r_n$ becomes separable and equal to

$$\left[\sum_{t_i=0}^{T/\Delta t} \alpha I(t_i) \Delta T \right]^n,$$

which tends to

$$\left[\alpha \int_t^{t+T} I(t') dt' \right]^n$$

as $\Delta t \to 0$. The remaining product can be expressed as follows:

$$\prod_{j=0}^{T/\Delta t} [1 - \alpha I(t_i) \Delta t] = 1 - \left[\sum_{i=0}^{T/\Delta t} \alpha I(t_i) \Delta t \right]$$

$$+ \frac{1}{2!} \left[\sum_{i=0}^{T/\Delta t} \alpha I(t_i) \Delta t \right]^2 - \frac{1}{2!} \sum_{i=0}^{T/\Delta t} \alpha^2 I^2(t_i) (\Delta t)^2$$

$$- \frac{1}{3!} \left[\sum_{j=0}^{T/\Delta t} \alpha I(t_i) \Delta t \right]^3 + \frac{1}{3!} \sum_j \sum_j \alpha^2 I(t_i) I^2(t_j) (\Delta t)^3$$

$$+ \cdots. \tag{4}$$

The terms in the square brackets are all of order zero in Δt, while the others are of the first and higher order in Δt and become negligible as $\Delta t \to 0$. It follows that

$$\prod_{i=0}^{T/\Delta t} [1 - \alpha I(t_i) \Delta t] \to \exp\left[-\sum_{i=0}^{T/\Delta t} \alpha I(t_i) \Delta t \right]$$

$$\to \exp\left[-\alpha \int_t^{t+T} I(t') dt' \right] \tag{5}$$

as $\Delta t \to 0$. Hence Eq. (1) gives Mandel's formula for photocount statistics, viz.,

$$p(n, t, T) = \frac{1}{n!} \left[\alpha \int_t^{t+T} I(t') dt' \right]^n \exp\left[-\alpha \int_t^{t+T} I(t') dt' \right]. \tag{6}$$

Equation (6) is evidently a Poisson distribution in n, with parameter $\alpha \int_t^{t+T} I(t') dt'$.

Appendix III

The degree of polarization of an electromagnetic Gaussian Schell-model source

The cross-spectral density matrix of a Gaussian Schell-model source is, according to Eqs. (5) and (6) of Section 9.4, given by the formula

$$W_{ij}^{(0)}(\rho_1, \rho_2, \omega) = \sqrt{S_i^{(0)}(\rho_1, \omega)} \sqrt{S_j^{(0)}(\rho_2, \omega)}\, \mu_{ij}^{(0)}(\rho_2 - \rho_1, \omega), \quad (i = x, y; j = x, y),$$

(1)

where the spectral density

$$S_i^{(0)}(\rho; \omega) = A_i^2 \exp[-\rho^2/(2\sigma_i^2)],$$

(2)

and the correlation coefficient

$$\mu_{ij}^{(0)}(\rho_2 - \rho_1; \omega) = B_{ij} \exp[-(\rho_2 - \rho_1)^2/(2\delta_{ij}^2)],$$

(3)

with the coefficients B_{ij} satisfying the constraints given by Eq. (7) of Section 9.4.

According to Eq. (14) of Section 9.2, the spectral degree of polarization of the electric field at a point specified by position vector ρ in the source plane is given by the general formula

$$\mathcal{P}^{(0)}(\rho, \omega) = \sqrt{1 - \frac{4 \operatorname{Det} \mathbf{W}(\rho, \rho, \omega)}{[\operatorname{Tr} \mathbf{W}(\rho, \rho, \omega)]^2}}.$$

(4)

When $\rho_2 = \rho_1 \equiv \rho$, Eq. (1) becomes

$$W_{ij}^{(0)}(\rho, \rho, \omega) = \sqrt{S_i^{(0)}(\rho, \omega)} \sqrt{S_j^{(0)}(\rho, \omega)}\, \mu_{ij}^{(0)}(0, \omega), \quad (i = x, y; j = x, y).$$

(5)

It follows that

$$\operatorname{Det} \mathbf{W}^{(0)}(\rho, \rho, \omega) = W_{xx}^{(0)} W_{yy}^{(0)} - W_{xy}^{(0)} W_{yx}^{(0)}$$

$$= S_x^{(0)}(\rho, \omega) S_y^{(0)}(\rho, \omega)[1 - |\mu_{xy}^{(0)}(0, \omega)|^2],$$

(6)

where we have used the fact that $\mu_{yx}^{(0)}(0, \omega) = \mu_{xy}^{(0)*}(0, \omega)$, the asterisk denoting the complex conjugate.

Further, it follows from Eq. (5) and from the fact that $\mu_{ii}(0, \omega) = 1$ ($i = x, y$) that

$$\text{Tr } \mathsf{W}(\rho, \rho, \omega) = S_x^{(0)}(\rho, \omega) + S_y^{(0)}(\rho, \omega). \tag{7}$$

On substituting from Eqs. (6) and (7) into Eq. (4) we obtain for the degree of polarization in the plane $z = 0$ the expression

$$\mathcal{P}^{(0)}(\rho, \omega) \equiv \sqrt{1 - \frac{4S_x^{(0)}(\rho, \omega)S_y^{(0)}(\rho, \omega)}{[S_x^{(0)}(\rho, \omega) + S_y^{(0)}(\rho, \omega)]^2}[1 - |\mu_{xy}^{(0)}(0, \omega)|^2]}. \tag{8}$$

On substituting from Eqs. (2) and (3) into this formula we obtain for the degree of polarization the expression

$$\mathcal{P}^{(0)}(\rho, \omega) = \sqrt{1 - \frac{4A_x^2 A_y^2 \exp\left[-\frac{1}{2}\rho^2\left(\frac{1}{\sigma_x^2} + \frac{1}{\sigma_y^2}\right)\right]}{\left[A_x^2 \exp\left(-\frac{\rho^2}{2\sigma_x^2}\right) + A_y^2 \exp\left(-\frac{\rho^2}{2\sigma_y^2}\right)\right]^2}(1 - |B_{xy}|^2)}. \tag{9}$$

We note that if

$$\sigma_y = \sigma_x \tag{10}$$

Eq. (9) reduces to

$$\mathcal{P}^{(0)}(\rho, \omega) = \sqrt{1 - \frac{4A_x^2 A_y^2}{(A_x^2 + A_y^2)^2}(1 - |B_{xy}|^2)}. \tag{11}$$

This formula implies that *when the condition (10) is satisfied*, i.e. when the r.m.s. widths of the spectral densities of the x and the y components of the electric field in the source plane $z = 0$ are the same, the degree of polarization $\mathcal{P}^{(0)}(\rho, \omega)$ is independent of ρ, i.e. *it is the same at every source point*. If, in addition, $|B_{xy}| = 1$ then, according to Eq. (3), $|\mu_{xy}(0, \omega)| = 1$, implying that *the x and the y components of the electric field are then completely correlated at each source point*. Equation (11) then gives $\mathcal{P}^{(0)}(\rho, \omega) = 1$, i.e. the degree of polarization is now unity at each source point, indicating that *the electric field is, in this case, completely polarized across the source*.

Appendix IV

Some important probability distributions

In this appendix we briefly consider some of the most important probability distributions and summarize their main properties.

(a) The binomial (or Bernoulli) distribution and some of its limiting cases

The binomial distribution, also known as Bernoulli's distribution, is one of the most important probability distributions encountered in classical physics. Some other well-known distributions such as the Poisson distribution and the Gaussian (or normal) distribution may be regarded as limiting cases of it.

The binomial distribution applies to experiments of the repetitive kind such as, for example, a succession of throws of dice. Let P be the probability of a certain event occurring in a single trial. Let us determine the probability $p_N(n, P)$, $(n = 0, 1, 2, \ldots, N)$, of exactly n successes occurring in N independent trials. We will denote by S a successful outcome and by F a failure. One possibility of having exactly n successes in N trials could be an outcome indicated by the sequence

$$\underbrace{S \quad S \quad S}_{n \text{ times}} \qquad \underbrace{F \quad F \quad F \quad F \quad F \quad F}_{N-n \text{ times}}. \tag{1}$$

By the so-called product rule for independent events[1] the probability of such an outcome is

$$\underbrace{P \times P \times P \ldots}_{n \text{ times}} \times \underbrace{Q \times Q \times Q \ldots}_{N-n \text{ times}} = P^n Q^{N-n} = P^n (1 - P)^{N-n}, \tag{2}$$

where $Q = 1 - P$ denotes the probability of a failure of the event to occur in a single trial.

We are actually not interested in the particular sequence (1) but rather in all sequences with n Ps and $(N - n)$ Qs, i.e. n successes and $N - n$ failures, *irrespective of the order of*

---

[1] See, for example, J. F. Kenney and E. S. Keeping, *Mathematics of Statistics, Part Two*, second edition (D. Van Nostrand, New York, 1951), p. 10.

the Ps and the Qs. The number of such sequences is evidently the same as the number of ways in which one can distribute n particles among N boxes. This number is equal to the *binomial coefficient*

$$^{N}C_n \equiv \binom{N}{n} = \frac{N!}{(N-n)!\,n!}. \tag{3}$$

Using Eqs. (2) and (3) we see that the probability of n successes in N trials is

$$p_N(n, P) = {}^{N}C_n P^n (1-P)^{N-n} \quad (0 \le n \le N). \tag{4}$$

This formula represents the *binomial distribution*.

It is not difficult to show that the first two moments of this distribution are

$$\bar{n} \equiv \sum_{n=0}^{N} n p_N(n, P) = NP, \tag{5a}$$

$$\overline{n^2} \equiv \sum_{n=0}^{N} n^2 p_N(n, P) = NP(Q + NP) \tag{5b}$$

and that the variance

$$\sigma^2 \equiv \overline{(n-\bar{n})^2} = NPQ. \tag{5c}$$

It may be shown that in the simultaneous limit of a very large number of trials ($N \to \infty$) and a very small probability of success in a single trial ($P \to 0$), subject to the constraint that the mean is still given by Eq. (5a), the binomial distribution given by Eq. (4) can be shown to become the Poisson distribution[1]

$$p(n) = \frac{\bar{n}^n e^{-\bar{n}}}{n!} \tag{6}$$

$\bar{n}$ being, of course, the mean of the distribution.

The Poisson distribution is found to be a good approximation to the binomial distribution when simultaneously $N \gg 1$ and $n \ll \bar{n}$.

One can readily verify that the second moment and the variance of the Poisson distribution are given by the expressions

$$\overline{n^2} = \bar{n} + \bar{n}^2, \tag{7a}$$

$$\sigma^2 = \bar{n}. \tag{7b}$$

[1] J. F. Keeney and E. S. Keeping, *Mathematics of Statistics, Part One,* third edition (D. van Nostrand, New York, (1954), p. 153.

Another important limiting case of the binomial distribution is the Gaussian distribution (also known as the normal distribution)

$$p(n) = \frac{1}{\sigma\sqrt{2\pi}} e^{-(n-\bar{n})^2/(2\sigma^2)}, \tag{8}$$

expressed here in terms of its mean $\bar{n}$ and its variance σ^2. It may be formally derived from the binomial distribution as $N \rightarrow \infty$, under certain constraints.[1]

(b) The Bose–Einstein distribution

As we noted earlier, the binomial coefficient $^N C_n$, defined by Eq. (3), can be understood from the analogy with the problem of distributing n particles among N boxes. The particles are regarded as physically distinct, i.e. they can be distinguished from each other.

Towards the end of Section 7.4 we encountered the concept of a cell of phase space, discussed in Appendix I. In such a domain, photons and some other particles are, as a consequence of Heisenberg's quantum-mechanical uncertainty principle, indistinguishable from each other. It is clear that under these circumstances the binomial distribution will not apply. Let us examine this situation.

The probability that n such particles are located in a cell of phase space is given by Eq. (2) as before. However, the binomial coefficient $^N C_n$, given by Eq. (3), is no longer appropriate because it represents the number of ways in which n *distinguishable* particles are distributed. In place of Eq. (4), one then has

$$p_N(n, P) = K_N P^n (1 - P)^{N-n}, \tag{9}$$

where the factor K_N has to be chosen so that the probability p_N is properly normalized. One might expect that this factor also depends on n, but more refined analysis shows that, in fact, this is not so.

Equation (9) has to be normalized so that

$$\sum_{n=0}^{N} p_N(n, P) = 1. \tag{10}$$

On substituting from Eq. (9) into Eq. (10) one readily finds that

$$K_N = \frac{1-\alpha}{1-\alpha^N} \frac{1}{Q^N}, \tag{11}$$

---

[1] See, for example, A. Papoulis and S. Unnikrishna Pillai, *Probability, Random Variables and Stochastic Processes*, fourth edition (McGraw-Hill, Boston, 2002), pp. 156–157, Theorem 5.2.

where $Q = 1 - P$ as before and

$$\alpha = \frac{P}{Q}. \tag{12}$$

On substituting from Eq. (11) into Eq. (9) we obtain for p_N the expression

$$p_N(n, P) = \frac{1 - \alpha}{1 - \alpha^N} \frac{1}{Q^N} P^n Q^{N-n} = \frac{1 - \alpha}{1 - \alpha^N} \alpha^n. \tag{13}$$

Suppose that the number of the particles is very large, i.e. that $N \gg 1$, and we formally proceed to the limit as $N \to \infty$. In order that the expression (13) tends to a finite limit we must evidently have $\alpha < 1$. The formula (13) then becomes

$$p(n) = (1 - \alpha)\alpha^n, \tag{14}$$

where we have now written $p(n)$ in place of $p_N(n, P)$. The formula (14) is one form of the so-called *Bose–Einstein distribution for one cell of phase space*.

Straightforward calculations give for the mean, for the second moment and for the variance of this distribution the expressions

$$\bar{n} \equiv \sum_{n=0}^{\infty} np(n) = \frac{\alpha}{1 - \alpha}, \tag{15a}$$

$$\overline{n^2} \equiv \sum_{n=0}^{\infty} n^2 p(n) = \bar{n}, \tag{15b}$$

$$\sigma^2 \equiv \overline{(n - \bar{n})^2} = \bar{n} + \bar{n}^2. \tag{15c}$$

Using the formula (15a) one may express the Eq. (14) in the form

$$p(n) = \frac{\bar{n}^n}{(\bar{n} + 1)^{n+1}}, \tag{16}$$

which is a more familiar form of the Bose–Einstein distribution for one cell of phase space.[1]

---

[1] The Bose–Einstein distribution for several cells of phase space is given, for example, in L. Mandel, *Proc. Phys. Soc.* (London), **74** (1959), 233–243.

Author index

Subject index

Printed in the United States
by Baker & Taylor Publisher Services